Dancing with Time

Cultural Interactions
Studies in the Relationship between the Arts

Edited by J.B. Bullen

Volume 43

PETER LANG
Oxford · Bern · Berlin · Bruxelles · New York · Wien

John Powell

Dancing with Time

The Garden as Art

PETER LANG
Oxford · Bern · Berlin · Bruxelles · New York · Wien

Bibliographic information published by Die Deutsche Nationalbibliothek.
Die Deutsche Nationalbibliothek lists this publication in the Deutsche Nationalbibliografie; detailed bibliographic data is available on the Internet at http://dnb.d-nb.de.

A catalogue record for this book is available from the British Library.

Library of Congress Cataloging-in-Publication Data:

Names: Powell, John, 1950- author.

Title: Dancing with time : the garden as art / John Powell.

Description: Oxford ; New York : Peter Lang, [2019] | Series: Cultural interactions: studies in the relationship between the arts ; 43 | Includes bibliographical references and index.

Identifiers: LCCN 2018057320 | ISBN 9781789971415 (alk. paper)

Subjects: LCSH: Gardens--Philosophy. | Art--Philosophy.

Classification: LCC SB454.3.P45 P69 2018 | DDC 635--dc23 LC record available at https://lccn.loc.gov/2018057320

Cover image: Topiary sundial in the garden in June at Ascott, Buckinghamshire. © National Trust Images/Colin Sturges.

ISSN 1662-0364

ISBN 978-1-78997-141-5 (print) • ISBN 978-1-78997-142-2 (ePDF)
ISBN 978-1-78997-143-9 (ePub) • ISBN 978-1-78997-144-6 (mobi)

Published by Peter Lang Ltd, International Academic Publishers,
52 St Giles, Oxford, OX1 3LU, United Kingdom
oxford@peterlang.com, www.peterlang.com

This publication has been peer reviewed.

In Memory of My Parents

Contents

Illustrations

Preface

I have called my book *Dancing with Time* because that title invokes three important, intertwined themes of the study: gardens change in time; we, in our own time, move and participate freely in gardens' times and spaces; and our garden encounters have the potential for artfulness. Time, movement, and sensory awareness are at the heart of the garden experience.

The book's subtitle, *The Garden as Art*, may suggest that two important questions – what is "art"? and, what sort of gardens may be works of art? – have been settled or, if not, are to be explored in the text. However, these and related questions have already been carefully debated by philosophers of art and my concern is not primarily with them. Let me, for the moment, proceed on the basis that you have at least a rough idea of what art is and what sort of gardens may be artful and later, I will introduce theories and insights, which vary from time to time and place to place, that philosophers have offered to confirm, and at times deny, that certain sorts of gardens can indeed be art, and to define what, after all, is "art." For now, suffice to say that the concept of "art" remains hotly debated today and that Western art gardens have enjoyed a chequered relationship with philosophical aesthetics, being, at different times, both lauded and rejected as exemplars of art and, for most of the last 150 or so years, largely ignored.

In this study, my concern is not with whether gardens can be art – for the most part, I assume that appropriate ones can be – but with how gardens, assuming they are art, are conceptualized within the arts and by philosophy. How are gardens the same as and different from the other arts? And, more importantly, what do any similarities and differences they possess imply for the ways in which we experience and understand gardens, and the ways in which we conceptualize their ontology, or, more simply, the ways in which they exist?

There exists already a literature devoted to examining the place of gardens in contemporary aesthetics. So, you may rightly ask, what is wrong with these existing accounts? Is a new one indeed needed? And, if a new account *is* needed, what features should it exhibit and what aspects of gardens should it consider? These questions are the guiding ones for this study, and answering them determines, broadly speaking, the book's shape and direction, and provides its motivation and momentum.

In an earlier study, Ismay Barwell and I invoked the American philosopher Susanne Langer's philosophy of art in order to account for gardens' *modus operandi* and, in that context, I proposed some similarities between gardens and music in terms of their temporal natures and their abilities to function rhythmically.[1] In this book, these "threads" of interest are reconsidered, interwoven with new elements, and their implied implications for the ways in which we experience gardens, and for gardens' unique ontological status, temporality, and modes of signifying meaning are drawn out. In particular, the idea of gardens' being temporal entities is developed further, and in new directions, and it assumes a central role in the study.

1 Ismay Barwell and John Powell, "Gardens, Music, and Time," in *Gardening: Cultivating Wisdom* (Oxford: Wiley-Blackwell, 2010).

Acknowledgements

During the writing of *Dancing with Time*, I have been fortunate to be able to draw on the substantial writings of the three key authors of the recent renaissance of philosophical writings on gardens. This renaissance began with the 1986 publication of a paper by Mara Miller, followed by further papers by Miller and, in 1993, her ground-breaking book, *The Garden as an Art*. The second key author, Stephanie Ross, published her foundational text, *What Gardens Mean*, in 1998, and the third author, David Cooper, published his important book, *A Philosophy of Gardens*, in 2006.[1] Although I disagree with many of the claims these philosophers make, I cannot underestimate the importance and value of their pioneering work for my project.

These three authors continue to contribute to debates in the field and they have been joined more recently by several others, two of whom have been of considerable value to me during the writing of this book. I have found the writings of Mateusz Salwa stimulating and helpful, even while I disagree with his claims regarding the "performance" of gardens.[2] And I have found Marta Tafalla's writings on the multi-sensual and, particularly, the olfactory experiences of gardens useful and illuminating.[3]

Outside the sphere of the philosophy of gardens, I am indebted to the philosophers Jerrold Levinson, Philip Alperson, and Curtis Carter, and the writer-architect Jan Birksted. Levinson and Alperson's "What Is a Temporal

1 David E. Cooper, *A Philosophy of Gardens* (Oxford: Oxford University Press, 2006); Mara Miller, *The Garden as an Art* (Albany: SUNY Press, 1993); "Gardens as Works of Art: The Problem of Uniqueness," *British Journal of Aesth*etics 26, no. 3 (Summer, 1986); Stephanie Ross, *What Gardens Mean* (Chicago: University of Chicago Press, 1998).

2 Mateusz Salwa, "The Garden as a Performance," *Proceedings of the European Society for Aesthetics* 5 (2013).

3 Marta Tafalla, "Anosmic Aesthetics," *Estetika: The Central European Journal of Aesthetics* 50, no. 1 (2013); "Smell and Anosmia in the Appreciation of Gardens," *Contemporary Aesthetics* 12 (2014).

Art?" furnishes the backdrop against which I have written Chapter 5 of the book. I am grateful to the authors for their detailed and closely reasoned investigation into temporal arts. Their careful groundwork has made my task of extending their criteria for temporal art so as to include gardens art that much easier than it would otherwise have been.[4] A serendipitous reading of Carter's "Improvisation in Dance" led, eventually, to my revising of the standard accounts of the ways in which we experience gardens.[5] And I was pleased to find aspects of this revision supported, if indirectly, in Birksted's important 2003 paper, "Landscape History and Theory: From Subject Matter to Analytic Tool."[6]

I am grateful to readers of earlier draft versions of materials for the book. In particular, I wish to thank Dr Sondra Bacharach of Victoria University of Wellington for her insightful comments and advice and encouragement; and Professor Samer Akkach of the University of Adelaide, without whom this book would, quite simply, never have been written. Professor Akkach's encouragement, wisdom, scholarship, and personal warmth have been more help than I can say. I would also like to acknowledge Professor Jenny McMahon of the University of Adelaide who gave me helpful advice at the initial stage of this project.

Finally, I am grateful to those at Peter Lang who have guided my manuscript through to publication. In particular, I would like to thank Dr Laurel Plapp, Michael Garvey, Simon Phillimore, and Jonathon Smith for their helpful criticisms and advice and their efficiency in dealing with all my requests.

4 Jerrold Levinson and Philip Alperson, "What Is a Temporal Art?," *Midwest Studies in Philosophy* 16 (1991).

5 Curtis L. Carter, "Improvisation in Dance," *The Journal of Aesthetics and Art Criticism* 58, no. 2 (2000).

6 Jan Kenneth Birksted, "Landscape History and Theory: From Subject Matter to Analytic Tool," *Landscape Review* 8, no. 2 (2003).

Introduction

I have written this book in order to offer alternatives to what I claim are some erroneous conceptions and arguments defended by contemporary analytic philosophy with respect to Western art gardens. I do not, of course, claim that all analytic philosophy's talk of art, and gardens is misguided and, certainly, I do not claim that philosophy needs new resources to understand adequately the phenomenon of gardens. In fact, the opposite is the case. Contemporary philosophy already has all the resources it needs to understand gardens and how we experience them. It is rather, as I hope to convince you, a matter of understanding and conceptualizing gardens and their place in the "artworld" better, more *appropriately*, in the first place and, only after that, seeking out philosophically valid ways of explaining accurately their ontology, experiential modes, and the unique nature of their principal materials (plants).

A New Account of Gardens

I offer a new account of art gardens that focuses on three aspects of them: their ontology, their experiential modes, and their materials. My account does not, therefore, match up to the substantial nature of accounts developed for some other individual arts. My aim is more modest. It is to develop, by focusing on these three key aspects, a framework within which to theorize about and conceptualize art gardens as a unique and uniquely valuable art form. Such a framework will also support the new understanding of the garden experience referred to in the book's title, *Dancing with Time*.

My account will allow gardens' distinctive qualities, features, processes, aesthetic possibilities, and arthood to be celebrated. Celebration of gardens' uniqueness has turned out to be difficult, even impossible, in the context of existing pan-art accounts of them, according to which gardens typically manifest as objects to be massaged or squeezed into pan-art conceptions. Rather than celebrating gardens' unique, distinctive qualities, such accounts have ended up treating those same qualities as if they were problems.

Although there exist voluminous historical and contemporary literatures devoted to philosophical aesthetics and also to a wide range of garden-related topics, it is only comparatively recently – since about 1990 in fact – that the literature related to the philosophical aesthetics of *gardens* has emerged from the quiescence into which it lapsed following its eighteenth-century heyday. Over the last 25 years, a slowly growing number of publications has appeared to investigate the aesthetics of gardens, and gardening, chief among which have been substantial monographs by Mara Miller, Stephanie Ross, and David Cooper.[1] These publications have treated the subject from a variety of points of view. The focuses of those dealing with art gardens have included gardens' temporality, modes of meaning, and historical and current status as works of art, the nature of the garden experience, and the definition of "art." However, it should not be understood from this renaissance of scholarly interest that gardens have only recently become more philosophically interesting and challenging and, consequently, in need of explication. Gardens have always been thus, but developments in philosophical aesthetics have made the task of that explication timely and, perhaps, more feasible than before.

Running through much of this recent literature is a pattern of assessing the art form of gardens through the lens of pan-art, one-size-fits-all definitions, theories, and ontologies of art and, in terms of gardens' meaningfulness and *modus operandi*, through the lenses of other arts, most notably painting and poetry. These approaches are all valid and offer interesting insights. However, there is a lack of a philosophical framework within which gardens' unique artfulness may be accepted as an irreducible given, and as a necessary starting point for an adequate consideration of gardens.

My aim of presenting an individual, artform specific account of the garden artform builds on a trend emerging in the writings of

1 Cooper, *A Philosophy of Gardens*; Miller, *The Garden as an Art*; Ross, *What Gardens Mean*.

twenty-first-century philosophers of art including, among others, Dominic McIver Lopes. Lopes writes that "there is nothing left to explain when you have [adequate] theories of the art forms and the arts."[2] I agree with him that the subject needs to be changed from art to arts, and artforms, and that such accounts are more informative and useful than pan-art accounts of the arts. In developing my own account of gardens, I draw from a range of sources, including the writer-architect Jan Birksted, and the philosophers Philip Alperson, Curtis Carter, and Arthur Danto. I also engage with some of the more accessible contemporary philosophical thinking on identity.

I have referred above to erroneous conceptions and arguments in contemporary aesthetics with respect to art gardens. This state of affairs has not appeared out of thin air. It has had a long genesis. Up until about 1800, gardens and aesthetics were closely linked in a mutually approving relationship. Gardens then gradually slipped beyond the view of aesthetics, for a variety of reasons, until the recent renaissance, at which time they have become of interest again, even if, in the view of Ross, they constitute a "dead" art.[3]

There is considerable historical and philosophical archival material available concerning garden art during the eighteenth and early nineteenth centuries and a comparative dearth of material for the remainder of the nineteenth (and twentieth and twenty-first) centuries. This paucity of material is unique to the art of garden making, the loss of status and consequent lack of historical and philosophical information of which has no equivalent in the other arts. A significant part of accounting for this changing importance and relevance of gardens to philosophical aesthetics involves investigating why the status of gardens changed from a high art in the eighteenth century to a non-art in the nineteenth century, and beyond. It is important to acknowledge this shift in status for two reasons. First, the changing philosophical attitudes to "art" and "nature" during the eighteenth and nineteenth centuries were highly influential in the contemporary changing status of gardens as art and the importance of

2 See: Dominic McIver Lopes, "Nobody Needs a Theory of Art," *The Journal of Philosophy* 105, no. 3 (2008), 127.

3 Ross, *What Gardens Mean*, 192.

this influence has remained largely unrecognized. And second, by clarifying why that phenomenon occurred, causes are in turn suggested for the twentieth century's philosophical disregard of gardens and, co-incidentally, the predominant focus of most professional landscape architecture away from gardens.

The majority of historical and contemporary writing on gardens has conceptualized them as poetic, pictorial, or architectural entities, which means that gardens are discussed as if they were poems, paintings, or buildings. While it is unreasonable to deny that garden do, to varying degrees, function in these ways, I aim to emphasize gardens' unique functional mode, which I claim is in good part temporal. In this study, I examine gardens' temporality closely and I compare manifestations of that temporality with manifestations of temporality in six non-garden arts.[4]

I employ the taxonomy of conditions for temporal arts developed by the philosophers Jerrold Levinson and Philip Alperson in their "What is a Temporal Art?" to determine whether gardens are theoretically temporal works of art.[5] In short, I find their account inadequate to the case of gardens and I propose a new condition to be added to their taxonomy to allow for the admission of gardens, and some other arts, to the category of "temporal arts." I claim that gardens' specific and unique temporal nature cannot be accommodated in pan-art notions of art, even in those that acknowledge art's broadly temporal nature, and therefore a unique and distinctive account of their temporality needs to be developed.

In the late twentieth and the twenty-first centuries, an acknowledgement of the importance of the temporality of gardens presents interesting opportunities for new ways of positioning gardens in the wider cultural field. Ironically, the emergence of *non*-art and *non*-garden aesthetics, such

4 For an earlier study of gardens' temporal characteristics, see: Barwell and Powell, "Gardens, Music, and Time."; John Powell, "What Is Temporal Art? A Persistent Question Revisited," *Contemporary Aesthetics* 13 (2015).

5 Levinson and Alperson, "What Is a Temporal Art?."

as the aesthetics of the environment, and the everyday, has implicitly facilitated the possibility of gardens' being considered as temporal entities.[6]

Any philosopher writing on the temporal nature of gardens inevitably and simultaneously raises issues to do with gardens' ontology and in this way ends up grappling with the legacy of Immanuel Kant's description of landscape gardening as "painting with nature's materials."[7] Kant's description reinforced a contemporary view of gardens as painterly, non-temporal works of art, and this view remained largely unchallenged until painting itself challenged aesthetics. In the denouement that followed cubism and its more controversial descendants, gardens disappeared altogether from philosophical aesthetics, just as conventional landscape painting disappeared as a "valid" art genre.

The twentieth and twenty-first centuries' avant-gardes have produced countless works that have challenged traditional ontological theories. In response to such works, significantly revised ontologies of art have emerged and these have, albeit unwittingly, come closer to being adequate to the case of gardens. In the study, I engage with ontologies developed by three contemporary philosophers of art – Stephen Davies, Richard Wollheim, and Gregory Currie – and I propose amendments and alterations which would, were they of a mind to consider them, make their theories more accommodating of gardens.[8] I also posit a straightforward account of human identity as a heuristic to better understand the ontology of gardens.

6 Some contemporary philosophers *are* addressing the importance of temporality. See: Mara Miller, "Time and Temporality in the Garden," in *Gardening: Cultivating Wisdom*, ed. Dan O'Brien (Oxford: Wiley-Blackwell, 2010); Salwa, "The Garden as a Performance."

7 Immanuel Kant, *Critique of Judgement*, trans. J. C. Meredith (New York: Oxford University Press, 2007), § 51.

8 Gregory Currie, *An Ontology of Art* (New York: St Martin's Press, 1989); Stephen Davies, *Philosophical Perspectives on Art* (Oxford: Oxford University Press, 2010); Richard Wollheim, *Art and Its Objects* (Cambridge: Cambridge University Press, 1980).

Some Key Terms

"Art", "aesthetics", and "garden" are key terms in this study. None of the following "definitions" defines those terms in the pure philosophical, or Socratic, sense of providing a set of necessary and sufficient conditions for the term. However, because the terms have such wide and varied currency, I need to clarify what I am intending to mean when I use them in this study.

Unless otherwise specified or made clear by the context, when I use "art" I am referring to any one or more of the various branches of creative activity, such as music, poetry, and dance, which are sometimes collectively referred to as the fine arts. However, even at this general level, it is necessary to distinguish between two significantly different historical uses of the term.

This distinction is made clearly, if somewhat dogmatically, by philosopher Ananda Coomaraswamy.[9] Coomaraswamy describes pre-Renaissance (and traditional Oriental) art-making as a "religious" activity whose function is, in the words of Aquinas, to "imitate Nature in her manner of operation."[10] On this view, an artist's function is not to represent reality but to use her skill (art) to invoke the ultimate reality, the Forms, from which, according to the Neo-Platonists and their medieval philosophical heirs, mundane reality and humanity derive their essence, meaning, and value.

By contrast, Coomaraswamy describes Renaissance and all post-Renaissance art as empty, materialistic – that is, concerned with materials instead of immaterial Forms – and all- too human. In this period, which extends into our own century, he claims that art objects are made not *per artem* but *pro arte*.[11] In other words, artists are no longer valued, nor are their products made or valued for what they can show and teach us about God and Ultimate Reality. Instead, their products are made and are to be

9 See: Ananda Coomaraswamy, "The Philosophy of Mediaeval and Oriental Art," in *Coomaraswamy*, ed. Roger Lipsey (Princeton, NJ and Surrey, UK: Princeton University Press, 1977); Vishwanath S. Naravane, *Ananda K. Coomaraswamy*, vol. 75, Twayne World Leader Series (Boston, MA: Twayne Publishers, 1978).

10 Coomaraswamy, "The Philosophy of Mediaeval and Oriental Art," 52.

11 Suggested translation: not "through art" but "for art" (ibid., 58).

valued for their own non-instrumental, aesthetic sakes. In this connection Coomaraswamy quotes approvingly these wonderfully apt lines of the art historian H. J. Spinden: "Then came the Renaissance. ... Man ceased to be part of the universe, and came down to earth."[12]

It is Coomaraswamy's post-Renaissance conception of art which underlies all my discussions of art in this study. The art with which I am concerned not only falls historically within this period but also, and more importantly, it is the art of known individuals – artists – producing artefacts for themselves, patrons, or more general "consumers," which products are of intrinsic as much as instrumental interest and value: art, one could say, for art's, artists', and art-lovers' sakes.

The terms "aesthetic" and "aesthetics" have their etymological roots in the word *aisthetikos*, which for the ancient Greeks denoted sentience, or sensitivity to any sensations received through the bodily senses.[13] (This original sense of the word plays a role in its contemporary antonym, "anaesthetics.") The modern word "aesthetics" was coined in the early eighteenth century by the German philosopher Alexander Baumgarten, who used it to denote knowledge obtained through the senses rather than the intellect and, especially, to refer to the sensory experience of beauty.

As the eighteenth century progressed, Baumgarten's meaning of "aesthetics" changed in two significant ways. First, the sensory experience of beauty was joined by the sensory experiences of the sublime and the picturesque as matters fit for aesthetics. Second, Baumgarten's original use of the term to apply to the sensory experience of beauty in nature *and* art changed so that by the early nineteenth century, and continuing for the following 130 or so years, "aesthetics" came to be used almost exclusively to describe the sensory perception of beauty and other qualities in works of art, not in nature. In effect, "aesthetics" became a de facto contemporary term to describe what in the twentieth century developed into the philosophy of art.

"Philosophy of art" emerged as a new term and discipline as nineteenth-century "aesthetics" became an increasingly inadequate concept to

12 Quoted in: ibid., 432.

13 "Aesthetic, N. and Adj.," *OED Online* (Oxford University Press, June 2016), <http://www.oed.com/view/Entry/3237>.

deal with troublesome activities and products of the avant-garde art world. Notorious artworks, such as Marcel Duchamp's *Fountain* and John Cage's *4'33"*, dealt body blows to the nineteenth-century notion equating art with beauty, or the sublime or picturesque, at the same time as they called into question the very category of "art" itself.[14]

In response to these challenges, the discipline of philosophy of art emerged. Its concerns ranged well beyond those it inherited from traditional aesthetics. The new discipline engaged, and indeed continues to engage, with art in all its geographical, cultural, and social variety, and it importantly and necessarily concerns itself with defining what art is and with ontological issues such as permanence, forgery, and authorship.

As the new discipline evolved, the old term, "aesthetics," became coopted to new uses. The traditional association of the term with art was overturned as new areas of aesthetic interest were studied. The two most important new areas are the aesthetics of the human or built environment, including the aesthetics of the everyday, and the aesthetics of the natural environment. The aesthetics of everyday life concerns itself with everyday places and activities, such as shopping malls, sport, housework, and even itching![15] The aesthetics of the natural environment concerns itself with the natural world and at first glance appears to mark a return to part of Baumgarten's original use of the term to denote "beauty in nature." However, "nature," and our complex relationships with and scientific understanding of it have altered so deeply since Baumgarten's time that any

14 Of course, the emergence of the avant-garde did not result in an immediate, clear move away from "aesthetics" and towards "philosophy of art" on the part of philosophers. Some notable aestheticians, including Croce and Collingwood, remained largely untouched by the avant-garde, either because they were uninteresed in it or because their work preceeded its principal manifestations. See: Gordon Graham, "Expressivism: Croce and Collingwood," in *The Routledge Companion to Aesthetics*, eds B. Gaut and D. McIver Lopes (Abingdon, Oxon: Routledge, 2005).

15 See: Sherri Irvin, "Scratching an Itch," *Journal of Aesthetics & Art Criticism* 66, no. 1 (2008).

perceived connections between the contemporary and his use of the term are largely illusory.[16]

Three contemporary philosophers have set out to define the term "garden." According to Cooper, the task of defining "garden" is unproblematic. He writes, "'Garden' is an entirely familiar term, and nearly every English speaker knows what it means. Pressed to say what I mean, my response would be 'The same as you who are pressing me mean by it – so you already know what I mean.'"[17] This definition, though adequate to Cooper's project, which focuses on the eudaimonic aspect of the experience of gardens, is too imprecise for my purposes, and in particular it is inadequate for underpinning the claims I make in later chapters of the study.

In their respective books, Ross and Miller both provide more useful "garden" definitions. I have argued elsewhere against the adoption of Ross's definition, and for the purposes of this study I adopt a modified version of Miller's definition.[18] Miller's definition has two clauses and incorporates three "conditions." She does not identify the conditions as such and, consequently, a reader is not to know whether any or all of them are necessary or sufficient. Miller states that a garden is "[1] any purposeful arrangement of natural objects (such as sand, water, plants, rocks, etc.) with exposure to the sky or open air, [2] in which form is not fully accounted for by purely practical considerations such as convenience."[19] I accept that part of her first clause which states that a garden is "any purposeful arrangement of natural objects (such as sand, water, plants, rocks, etc.)."[20] It adequately identifies what a garden is even though it fails to point out some of their essential features, such as their inherent dynamism. But I do not accept the second part of her first clause because not all gardens need to have exposure

16 For a discussion of this point, see: Glenn Parsons, *Bloomsbury Aesthetics: Aesthetics and Nature* (London: Continuum, 2008).

17 Cooper, *A Philosophy of Gardens*, 13.

18 John Powell, "We Do Not Have an Adequate Conception of Art until We Have One That Accommodates Gardens" (MA, Victoria University of Wellington, 2012), 8–11.

19 Miller, *The Garden as an Art*, 15.

20 Along with Miller, I acknowledge that many gardens also incorporate unnatural objects, such as fibreglass sculptures, ceramic birdbaths, glass, mosaics, etc.

to sky or open air. Interior gardens, at scales ranging from Victorian terrariums to large-scale municipal conservatories, and "winter gardens" are examples of types of gardens that do not have exposure to sky or open air but which I believe should not thereby be excluded by any adequate definition of gardens.[21]

I reject the second clause of Miller's definition – "in which form is not fully accounted for by purely practical considerations such as convenience" – for two reasons. First, there is no good reason to deny that kitchen and herb gardens are gardens. Further, it is appropriate to take aesthetic pleasure in, and make aesthetic judgements about them. Second, gardens laid out according to practical considerations, such as ease of maintenance or optimum productivity, may be beautiful because they are well suited to a purpose.[22]

My interest is primarily in "art gardens" and Miller's shortened definition comfortably includes the gardens that are commonly considered to be works of art. Examples of such gardens are shown in Figures 1–6. They include: Monet's garden at Giverny; Ian Hamilton Finlay's garden, *Little Sparta*, at Dunsrye; Le Nôtre's gardens for the Palace of Versailles outside Paris; Henry Hoare's garden, Stourhead, in Wiltshire; Robert Irwin's gardens at the Getty Centre in Los Angeles; and the Australian Garden near Melbourne, designed by Taylor Cullity Lethlean and Paul Thompson. These gardens are all well known and highly regarded, and, like Miller, I am interested in them, their materials, and the arrangement of those materials.[23]

21 Some spectacular examples of gardens without exposure to the sky can be found at Singapore's Gardens By The Bay. (See: <http://www.gardensbythebay.com.sg/en.html>.)

22 In Kantian terms, gardens may exhibit free and dependent beauty. That is, they may be aesthetically pleasing both because they are "lovely to look at" and because they are "fit for purpose."

23 In this study, I have chosen not to examine the practice of placemaking, often carried by landscape architects and other designers professionals. The distinctions between gardens and "made" places are difficult to draw, and somewhat arbitrary. I have made the decision to concentrate on the former because of their historical and ongoing relationships with the fine arts and the philosophy of art.

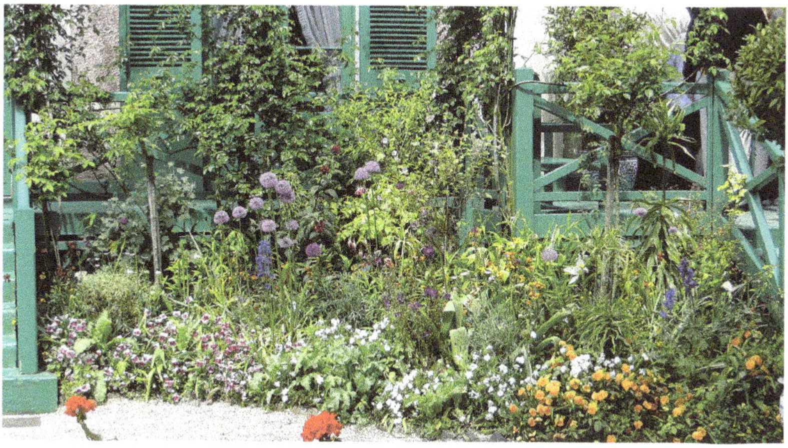

Figure 1. Photo of part of Monet's garden near Giverny, France. Reproduction of "Monet's garden in May (4)" by Richard White is licensed under CC BY 2.0.

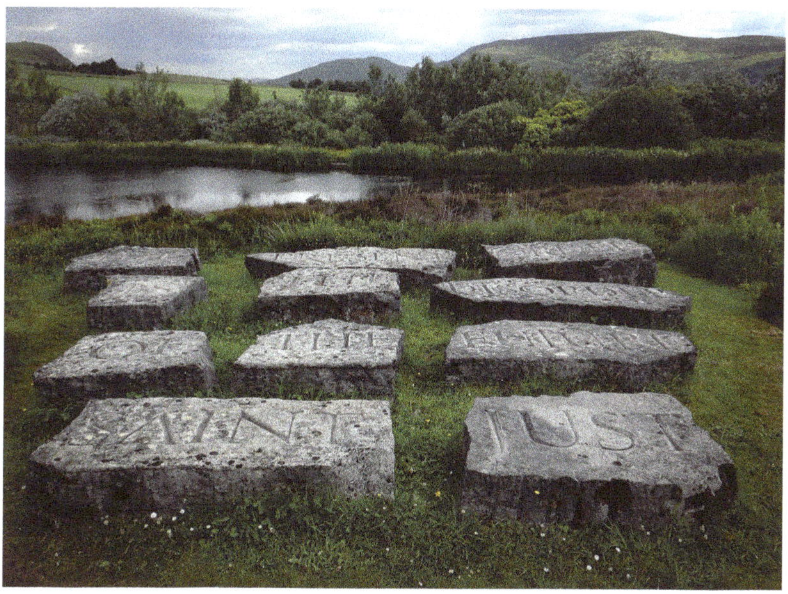

Figure 2. Photo of part of Ian Hamilton Finlay's garden, Little Sparta, at Dunsrye, Scotland. Reproduction of "1983, Ian Hamilton Finlay's Little Sparta garden: The Present Order is the Disorder of the Future" by Rosa Menkman is licensed under CC BY 2.0.

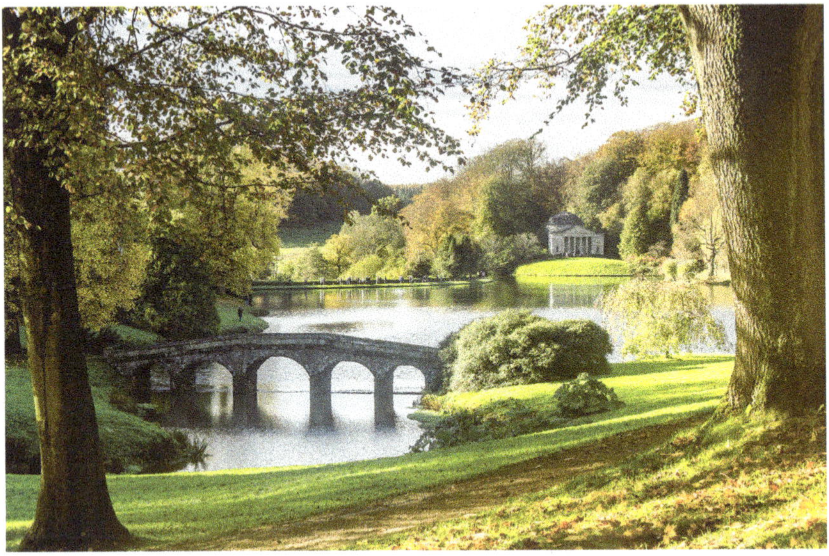

Figure 3. Photo of part of Henry Hoare's garden, Stourhead, in Wiltshire, England. Reproduction of "Stourhead Autumn" by Graham Duerden is licensed under CC BY 2.0.

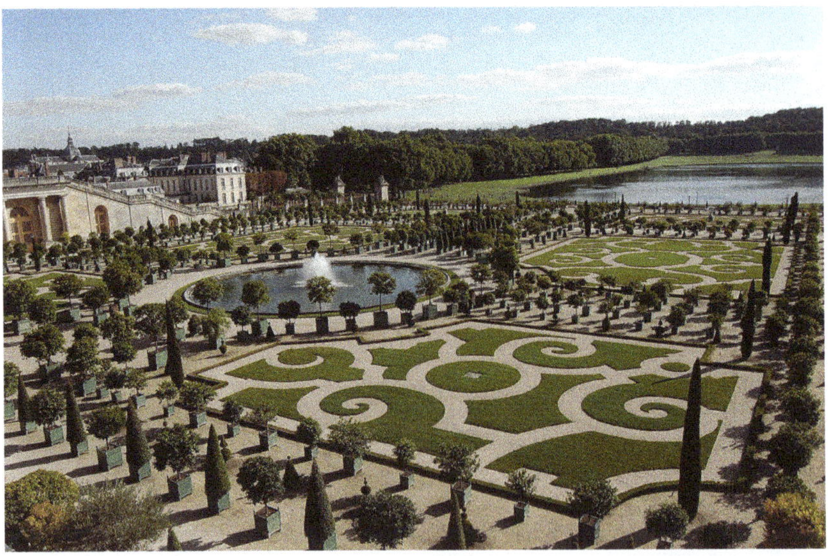

Figure 4. Photo of part of Le Nôtre's gardens for the Palace of Versailles, France. Reproduction of "Orangerie" by Emily Mocarski is licensed under CC BY 2.0.

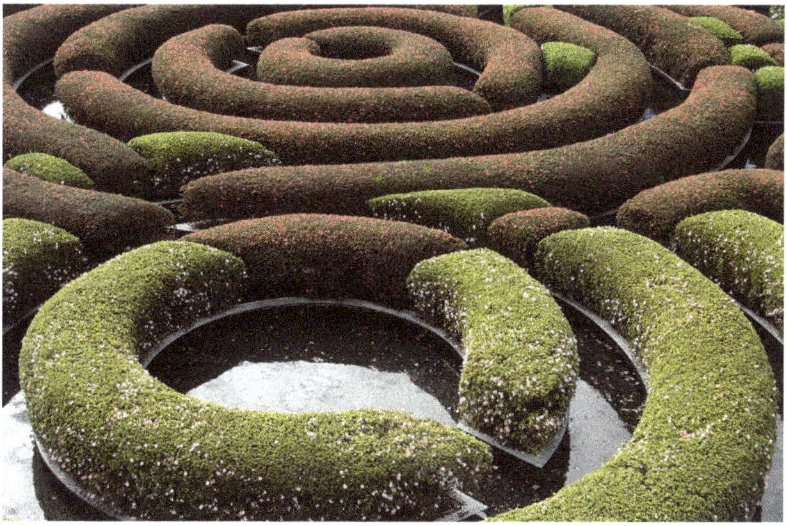

Figure 5. Photo of part of Robert Irwin's gardens for the Getty Centre, Los Angeles, CA, USA. Reproduction of "Getty Museum" by Ian Sanderson is licensed under CC BY 2.0.

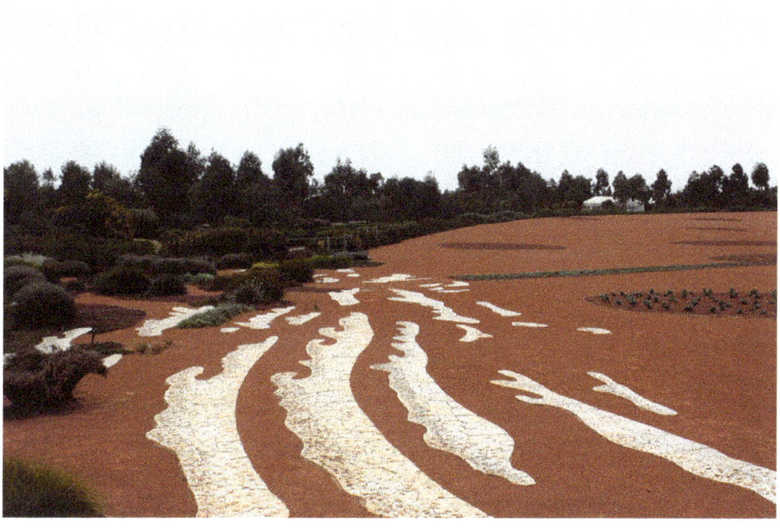

Figure 6. Photo of part of the Australian Garden, near Melbourne, Australia, designed by Taylor Cullity Lethlean and Paul Thompson. Reproduction of "Australian Desert Garden" by Greenstone Girl is licensed under CC BY 2.0.

Outline

The book is divided into four parts. The first part – "Setting the Scene" – comprises Chapter 1, in which I prepare the ground for the discussions in the following part, extrapolating from the case of an actual garden to elaborate on the difficulties that the existence and experience of gardens have posed, and continue to pose, for aesthetics.

The second part – "The Territory: A Survey" – comprises Chapters 2 and 3. The chapters offer an historical-cultural-philosophical survey of aesthetics, art, and gardens during the Enlightenment, the romantic age, and the periods of modernism and postmodernism. The survey provides an historical and conceptual context for the following part.

Part III – "The Lie of the Land: A Critique" – comprises Chapters 4, 5, and 6. It considers the adequacy of contemporary philosophy of art to the case of gardens. In particular, it critiques recent philosophical writing on gardens, assesses the adequacy of current theories of temporality to gardens and other arts, and evaluates contemporary theories of ontology with respect to their ability to account for gardens.

The fourth part – "Breaking New Ground" – comprises Chapters 7 and 8. It endorses the idea that artform-specific theories and ontologies of art may be more useful than pan-art ones and develops such an account of gardens, which accommodates their unique ontology, materials, and mode of experience.

PART I

Setting the Scene

CHAPTER 1

"But Is It Art?": Case Study of a New Zealand Garden

Gardens have unique and distinctive characteristics which philosophical aesthetics has typically misconstrued, or simply ignored. These characteristics may fairly be counted among gardens' most distinctive, interesting, and valuable ones. These characteristics are also ones which have placed gardens outside the category of what has been historically defined as, or defined as acceptable for, art. Tupare is a garden that exemplifies many of these characteristics. It is also, in many important ways, a typical garden. Examining Tupare will allow me to extrapolate my findings for that garden and show how, why, when, and for whom gardens in general are problematic art objects. In other words, Tupare will exemplify a problematics of gardens.

Tupare

Tupare is a 3.6 hectare garden located in New Plymouth, on the west coast of the North Island of New Zealand.[1] It is one of several exceptional gardens in a geographical area that possesses a climate and soils well suited to many styles of gardening. The garden was developed by the Matthews family and remained in its ownership for 50 years, until ownership was transferred to

1 "Tupare" is a Maori word meaning "garland of flowers." For a brief overview of the garden see: Taranaki Regional Council, "Tupare: Relive the Splendour," Taranaki Regional Council, <http://www.trc.govt.nz/tupare-home/>.

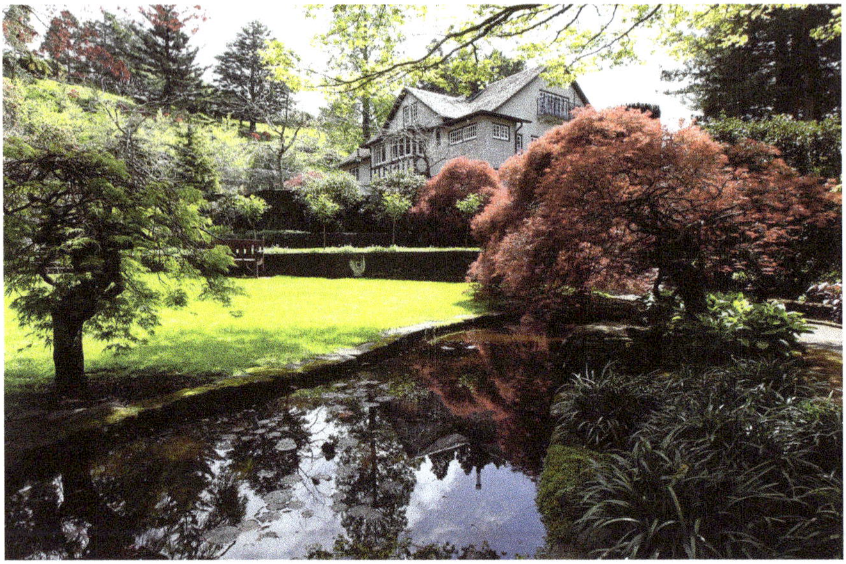

Figure 7. Photo of part of Tupare garden, near the house. Photo reproduced with permission from Rob Tucker.

the Queen Elizabeth II Trust in 1983 and then to the Taranaki Regional Council in 2002 (see Figure 7).

Its site is in many ways unlikely for a garden. It enjoys a pleasing aspect towards a river on part of its north-east and south-east boundaries, but the site's predominant topographical characteristic is its extremely steep slopes up to the north, west, and south-west boundaries, which, in southern hemisphere terms, makes the site far from ideal for gardening. The flatter areas of the site are occupied by a house, auxiliary buildings, and tennis court and, some 22 metres below the house site, by an area of un-gardened river flats (see Figures 8 and 9).

The evolution of Tupare has not been examined in any depth. However, it seems reasonable to assume that the garden developed its characteristics and flavour as a result of four influences. First, Sir Russell Matthews was a keen plantsman with an eye for the new and the fashionable. Second, he was a successful roading contractor who was able to command the services of his workers during the off season to construct paths, driveways, walls,

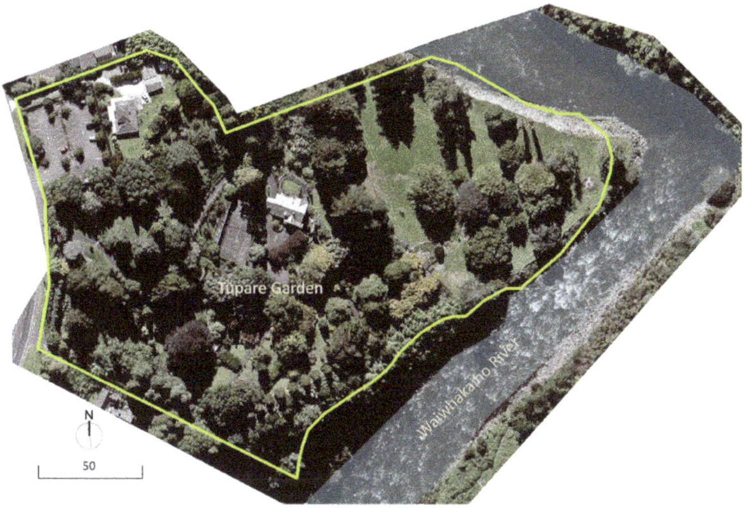

Figure 8. Aerial photo of Tupare garden showing location of the house, garden areas, ungardened river flats, and the Waiwhakaiho River along part of the boundary. Reproduction of "Aerial photo" by AAM NZ Ltd is licensed under CC BY 4.0.

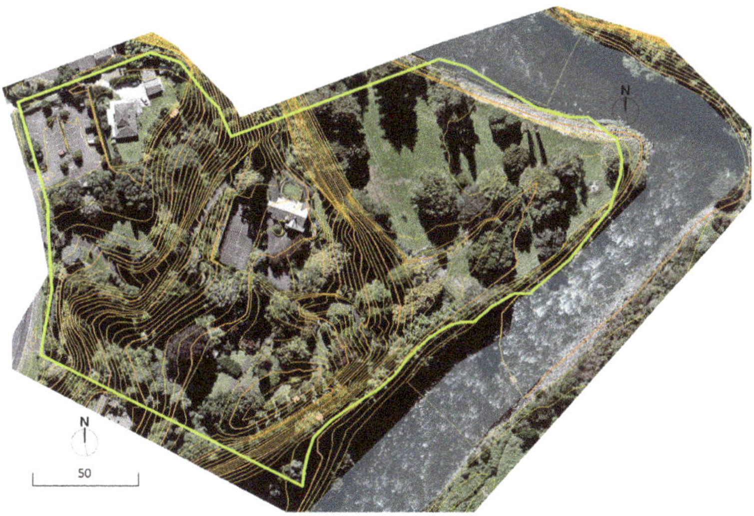

Figure 9. Aerial photo of Tupare garden, showing contours at 2 m intervals. Reproduction of "Aerial photo and contours" by AAM NZ Ltd is licensed under CC BY 4.0.

and so on, to make pedestrian and vehicular movement possible around an otherwise inaccessible site. Third, like many of his contemporaries, his social ambitions were founded in a preference for a certain type of Englishness. And fourth, his "English" house, designed in the arts and craft style by the noted architect J. W. Chapman-Taylor, required an appropriate, gardenesque garden to set it off.

Any attempt at describing or re-presenting Tupare in words and static images must fail to provide an adequate account of the garden and the range of sensory experiences it offers. Although this limitation applies to descriptions of all gardens, in the case of relatively unchanging, formal gardens on flat terrain, such description or re-presentation may be more or less successful, although still inadequate. However, in the case of highly changeable, highly sensual gardens sited on dramatically sloping topography, such as Tupare, such descriptions and re-presentations are not only inadequate, they may be misleading. So, with that necessary ekphrastic caution firmly in mind, let us begin.

Tupare is almost overwhelmingly multi-sensual in its appeal. It is rich in opportunities for tactile, olfactory, visual, aural, and kinaesthetic experiences, and, at least in the kitchen garden, for actual or promissory gustatory experiences. I have singled out for comment three categories of sensory experiences that are particularly strongly present in the garden: the aural, and what I think of as "pure," and "mixed" kinaesthetic experiences.

Tupare shares the usual range of opportunities for aural experiences offered to varying degrees by most gardens. At Tupare, you can hear birds' singing, listen to the sounds of wind in the trees, and hear the sounds of your own and other visitors' progress around the garden. But Tupare offers a further, unusual sonic treat. It is permeated by the aural presence of the Waiwhakaiho River, which forms part of its boundary (see Figure 8). The Waiwhakaiho's riverbed is strewn with boulders and this gives the fast flowing, shallow river a distinctive, noisy presence (see Figure 10).

Although the river cannot be seen from most of the garden, its sound spreads everywhere. A sensitive visitor remains aware of it even when she is looking at smaller water features within the garden. The river's sound functions somewhat in the nature of a sonic "view" – let us call it a hearing – around which the garden is assembled in the manner of a garden

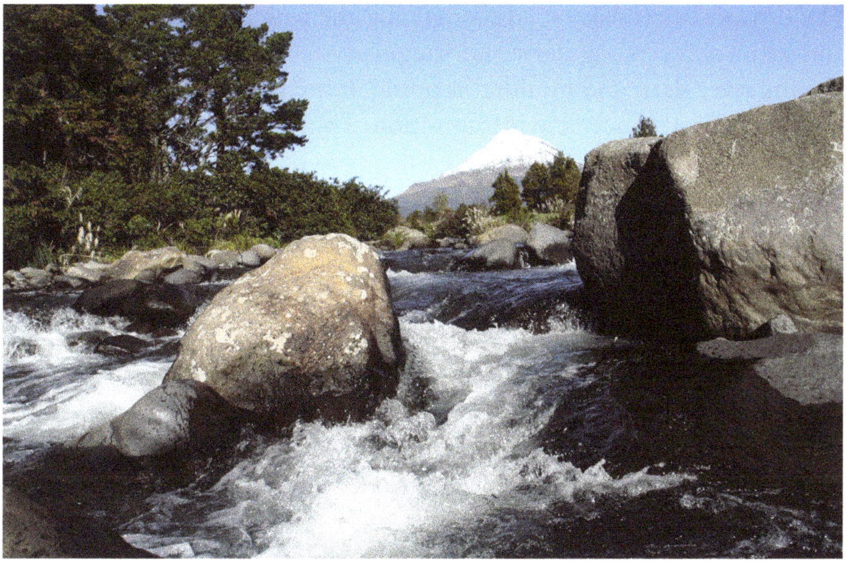

Figure 10. Photo of Waiwhakaiho River and its source, Mt Egmont / Mt Taranaki, taken from near Tupare garden. Photo reproduced with permission from Massey University, New Zealand.

assembled around an external (visual) view, and the garden's topography plays its role in amplifying or diminishing, and containing or excluding the sound. This phenomenon is, besides, a reminder of the permeability and indistinctness of gardens' boundaries, which characteristics are usually assumed to have visual implications only.[2]

I think of *pure* kinaesthetic sensation as the experience of the sensations provided by the body's proprioceptors, which are the sensory organs that provide information about the movement and position of the body.

2 Landscape architect Michael Fowler, adopting the terminology of the Canadian accoustic ecologist R. Murray Schafer, has introduced the apposite term "keynote" to describe a pervasive audible presence such as that of the Waiwhakaiho River at Tupare. See: Michael Fowler, "Soundscape as a Design Strategy for Landscape Architectural Praxis," *Design Studies* 34, no. 1 (2012), 3.

On Tupare's extremely varied topography, the sensations on offer are rich and various. Steep, regularly and irregularly ramped paths, steps, gently sloping river flats, and boulder hopping in the river provide the visitor with an unusual variety and juxtaposition of pure kinaesthetic experiences in a relatively small area, and the garden is to be valued on this account (see Figure 11).

However, the opportunities for *mixed* kinaesthetic experiences afforded by the garden are more valuable still. The contemporary writer-architect Jan Birksted describes a garden visitor, or, to use his term, beholder, as being "(extra)visual and mobile ... in gardens' three-dimensional space and time."[3] In the context of Tupare, this means that besides changes in the garden's living elements (plants), and other changes attributable to weather and season, a visitor's location is itself constantly changing in time and space. Add to this Tupare's topographical variety, and the result for the visitor is a sensory experience that is almost impossibly rich in potential. For example, a view from a narrow, winding, sloping path may be obliterated at a single step by the uppermost leaves of a tree growing from four metres below the level of the path, only to be revealed two steps later once the treetop is passed; during a short walk the same tree may be experienced at its base, middle and top; the scent and sight of flowers growing at the top of a tall tree may be available, at some times of the year, from a path high above the tree's base; a tree's weeping branches may descend on the viewer's path from a tree planted higher up the slope, allowing tactile and close visual and olfactory examination of, say, a conifer's unusual cone (see Figure 12). All such experiences may be expected and found in many gardens, but they exist in peculiar abundance on the hillsides of Tupare.

The question of the authorship of Tupare is both simple and complex. It is simple in that the garden is the vision of Sir Russell Matthews, who bought the land, commissioned and fought with the architect of the house, and caused the garden to be laid out and planted. Therefore, it is his garden. However, leaving aside the complications of Lady Matthews' almost certain involvement, and the recorded, enforced involvement of

3 Jan Kenneth Birksted, "Landscape History and Theory: From Subject Matter to Analytic Tool," *Landscape Review* 8, no. 2 (2003), 6.

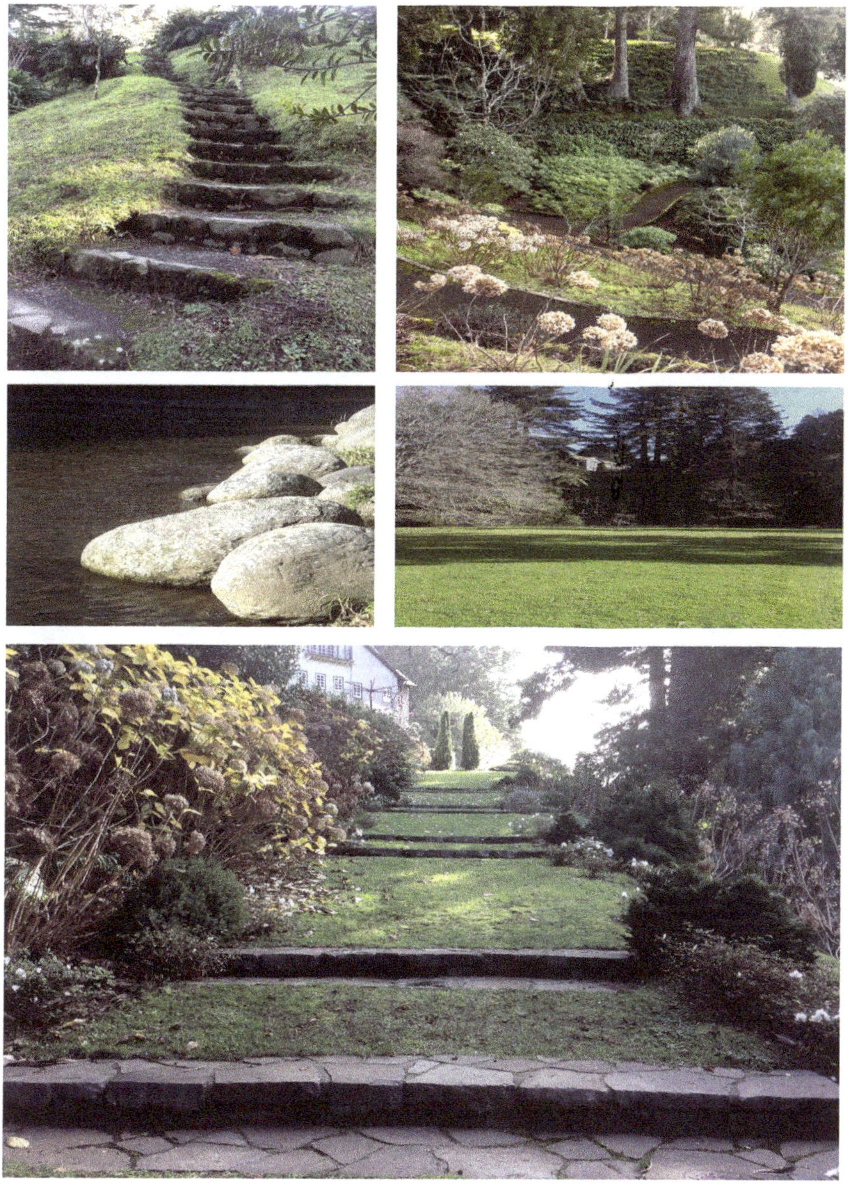

Figure 11. Composite image of paths, steps, river flats, and river boulders at Tupare. Author's photos.

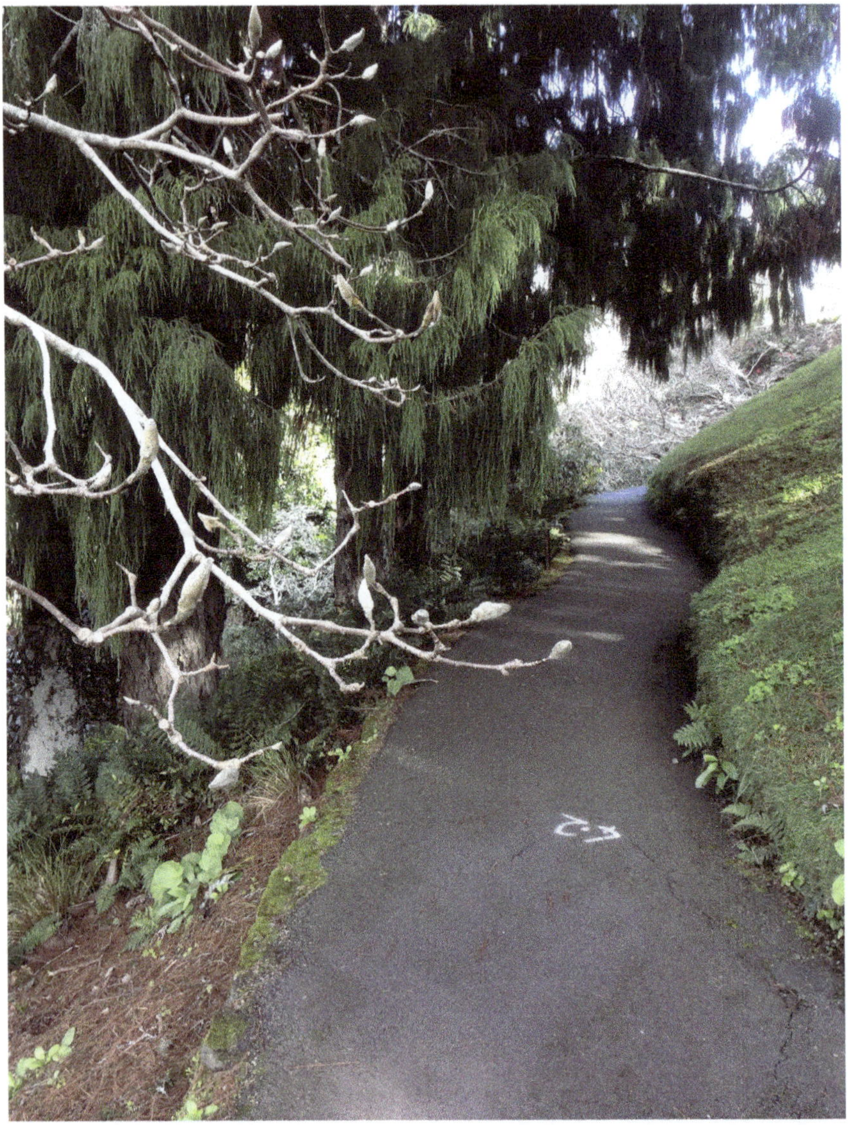

Figure 12. Photo of a magnolia and rimu growing from a lower level and overhanging a higher-level path at Tupare. Author's photo.

his children, what are we to make of the indispensable contributions made to the garden by Sir Russell's seconded road-making employees?[4] To what extent were they directed, or were they relatively free agents? No records exist to clarify this. And what was the role of the Queen Elizabeth II Trust in influencing the evolution of the garden during its 19-year tenure? Since 2002, under the ownership of the Taranaki Regional Council, it is clear that significant changes have been and continue to be made. For example, trees have been felled, replacements and other trees have been planted, and some complex, labour intensive underplantings have been replaced by more easily maintained mass plantings of ground cover plant species. So, in other words, while Russell Matthews' role as the original "author," or originating artist, of the garden is unlikely to be challenged, Tupare exemplifies a pattern of continual (re-)creation, maintenance, and replanting, carried out by a range of contributors and subsequent authors, that is typical of all but the shortest lived and simplest gardens.

I have so far treated Tupare as if it were firmly established in its arthood, but such is not the case. Rather, Tupare is an example of the sort of garden which Mara Miller and Stephanie Ross, each invoking philosopher George Dickie's definition of art, would seek to describe as non-art.[5] It suffers from many of the disadvantages that traditional gardens continued to exhibit in the face of the preferences, if not requirements, for art that were espoused by practitioners, theorists, apologists, and philosophers associated with the major art movements of the latter part of the twentieth century. Instead of being challenging, densely intellectual, and egalitarian, Tupare is beautiful and sensuous, and smacks of private wealth and power. Furthermore, in a city with an exceptionally healthy, progressive (visual) arts community and a noted public gallery, Tupare has found no place as

4 For descriptions of the family's and others' contributions to the evolution and maintenance of the garden see: Elizabeth M. Benney, "Tupare," in *The New Zealander's Garden*, eds Julian Matthews and Gordon Collier (Auckland, NZ: Endeavour Press, 1985), 27–9; Richard Matthews, *To Whom It May Concern: A Romp through the Life of Richard Matthews* (New Zealand: R. Matthews, 2009), 15.

5 In Chapter 3, I examine and reject Miller's and Ross's claims in this regard.

"art" – and perhaps not least because it cannot be exhibited in that, or indeed any, gallery.[6]

Emerging from the examination of Tupare are three elements which I believe are centrally relevant to a proper understanding and appreciation of gardens and with which, therefore, the philosophical literature ought to engage. They are gardens' ontology, the experience of gardens, and gardens' arthood and the artistic status of their makers. Let us see how Tupare measures up in the context of these three elements; and how standard accounts of these elements accommodate Tupare.

Standard ontological accounts of gardens have concentrated almost overwhelmingly on the visual dimension of gardens.[7] Such accounts have also usually ignored the permanent mutability of all garden elements or, at best, they have acknowledged major seasonal changes only. My description of Tupare has highlighted not only the changeability of that garden's visual elements but also, and more importantly, it has drawn attention to that garden's non-visual, and also changing, elements, including sound and scents. By failing to account for gardens' life-long, multi-sensual, multi-scaled mutability, standard ontologies of art fail to accommodate Tupare and, in fact, most other gardens too. In addition, such ontologies render needlessly problematic the retention of identify through change that is characteristic of Tupare and other gardens.

My necessarily selective account of the experience of Tupare highlighted that garden's potential for temporally open-ended, multi-sensual, and, in particular, kinaesthetic experiences. Historically, experiential accounts of gardens have typically failed to acknowledge or account for such phenomenological riches and complexities, which is perhaps unsurprising given the related ontological lacuna mentioned earlier, and they

6 The gallery referred to is the Govett-Brester Art Gallery, incorporating the Len Lye Centre. See: <http://www.govettbrewster.com/len-lye/>.

7 For two notable exceptions, see: Michael Fowler, "Sound as a Considered Design Parameter in the Japanese Garden," *Studies in the History of Gardens & Designed Landscapes: An International Quarterly* (2015); Tafalla, "Smell and Anosmia in the Appreciation of Gardens."

therefore remain inadequate to the case of Tupare. Further, the complex interrelations between the temporal nature(s) of a garden's materials and of the process of a visitor's experience of any garden has remained outside the scope of historical experiential accounts of art.

I have described Tupare as a beautiful, artistic garden, whose existence is attributable in large part to the vision, energy, and resources of Sir Russell Matthews. I have also noted that it is not recognized as a work of art by what might be called the professional, local "artworld" and that, according to standard accounts of artistic authorship, the role of Russell Matthews, and, indeed, any garden maker, *qua* artist, is open to scrutiny. In other words, philosophy's standard answers to the two related questions – what is a work of art? And who is the artist of this work? – are unfavourable to the case of Tupare and its maker.

The classification of Tupare as art, or otherwise, depends, of course, on the prior definition of art adopted by the classifier. Gardens have been classified as "art" from time to time during their long history. However, the twentieth century's definers of art did not have gardens in mind when formulating their definitions. Or, rather, they may have had, but the application of the definitions generated during the late twentieth century, such as those of George Dickie and Monroe Beardsley, have been used by other philosophers to exclude gardens from the category of art.

The question of authorship, or "artistship," has two aspects. First, a distinction needs to be made between the originator-artist of a garden and his contributors, and, second, there is the question of the artist status of a garden's originator. In the first case, Matthews is the artist of Tupare while his family, gardeners, and other contractors are contributors; just as, say, set painters, lighting technicians, make-up artists, costume designers, dancers, and musicians contribute to the artist-choreographer's ballet. These relationships have been acknowledged by philosophers of art. However, the intricacies of their interrelationships have not been adequately explored, especially in the case of gardens. In gardens, the matter is complicated because maintenance, and, in the case of Tupare, changing ownership, in conjunction with gardens' continuous mutability, may result in the "same" garden changing quite radically over a period. Philosophical aesthetics appears to accept that this is the case but does not delve too deeply into why or how this may be acceptable when, in other traditional arts, it is not.

In the case of the "artist" status of a garden's originator, it should be noted that the nineteenth-century's preference for a solitary-genius-artist cast a long shadow deep into the twentieth century.[8] And even today, "artist," when used in folk parlance and/or without further qualification, conjures up aspects of the nineteenth-century conception of an artist. Clearly Russell Matthews, the first and most prolific layer of macadam in New Zealand, and a highly practical man to boot, did not fit that mould, and may well have been embarrassed by attempts to style him "artist."[9]

Some Consequences for Philosophical Aesthetics

It *appears* that Tupare is not easily accounted for by existing accounts of the ontology, and experience of artworks, nor do it and its artist *appear* to be easily accommodated within current definitions of art and notions of authorship (or, "artistship"). Aesthetics, as traditionally practised, has largely failed, or chosen not, to engage with these aspects of art gardens. However, I have italicized "*appears*" above because I contend that philosophical aesthetics does indeed have the resources to deal effectively with gardens. The fact is that aesthetics has largely turned away from gardens, even while it has pursued with some passion new arts whose ontological, experiential, and arthood features and aspects gardens have exemplified

8 I discuss further the nineteenth-century's concept of "artist," and its implications for gardens and their designers, at 66.

9 Certainly, being practical and an excellent businessperson does not rule out the possibility of one's being an artist, as this recent book highlighting Capability Brown's strengths in those areas shows: David Brown and Tom Williamson, *Lancelot Brown and the Capability Men: Landscape Revolution in Eighteenth-Century England* (London: Reaktion, 2016). However, those strengths do tend to run counter to the nineteenth-century's preference for an artist's being a solitary, genius-seer.

for centuries. Examining how these features and aspects of gardens have fared in the context of philosophical aesthetics is an important focus of the rest of the book. However, at this point, and with their exemplification at Tupare in mind, I will comment briefly on how each of these features and aspects has been treated, or ignored, in the current literature.

The ontology of gardens concerns the ways in which gardens exist and persist, and the nature(s) of their physical materials. Gardens' ontologies are necessarily intertwined with their experiential and temporal dimensions and, perhaps for this reason, much ontological writing on gardens has concentrated on these dimensions. The subject of gardens' continuously changing material composition, and visitors' experiences of it, has been widely canvassed. For the garden historian John Dixon Hunt, gardens are about process and change. They are never finished, and nor do they even "allow the illusion of a stable and coherent object of study."[10] Cooper writes that, "it is in this relationship [between gardens and temporality and ephemerality] that the distinctiveness of garden aesthetics lies."[11] And Miller agrees that a distinctive aspect of gardens' ontology lies in their essential changeability.[12]

There have also been some investigations into the important ontological questions of whether gardens-*qua*-art are physical-, mental-, or action-type entities, and whether the worlds we explore in them are real or virtual. Both Miller and Ross make differing claims for gardens' being virtual worlds, and Hunt contrasts the inescapable physicality of gardens with what he calls their metaphysical aspect.[13]

Once gardens' ontological realm has been extended beyond the visual, as I have proposed is appropriate in the case of Tupare, and especially once

10 John Dixon Hunt, "Gardens: Historical Overview," in *Encyclopedia of Aesthetics*, ed. Michael Kelly (New York, Oxford: Oxford University Press, 1998), 272.

11 David E. Cooper, "Foreword," in *Gardening: Cultivating Wisdom. Philosophy for Everyone*, ed. Dan O'Brien (2010), ix.

12 See, for example: Miller, *The Garden as an Art*, 38–49; "Time and Temporality in the Garden," 178–91.

13 See: Hunt, "Gardens: Historical Overview," 272; Miller, *The Garden as an Art*, 127–31; Ross, *What Gardens Mean*, 175–88.

the mutability of gardens' materials has been recognized, then a range of epistemological questions present themselves, the underlying one of which is: what sort of artworks are gardens? Are they like poems and paintings, as Ross suggests?[14] Are they like music, dance, or drama, as the Polish philosopher Mateusz Salwa and others have proposed? Or are they a type of building or mobile sculpture?[15] These questions relate to how we experience gardens, and, in their turn they imply further questions, such as, what senses do (ought) we exercise in an appropriate experience of a garden?[16] Are gardens meaningful in the way of poems or paintings or something else? Are gardens performances? Should we experience gardens in preordained sequences or should visitors just wander about as they fancy? Do gardens include the view of the church spire or petrol station that is clearly framed between the trees? Where and when are a garden's spatial and temporal boundaries?

I claim that gardens' temporal qualities constitute a uniquely important aspect of them and provide an important reason for our valuing them. Gardens' temporality is, of course, an important aspect of their ontology, but it also is powerfully involved in our *experiences* of gardens. Cooper writes:

> For it is in this relationship [between gardens and temporality and ephemerality] that the distinctiveness of garden aesthetics lies – in the manner, say, that a garden "presents," or makes mindful of, time, or perhaps in an "enchantment" that unexpected changes in the process of experiencing a garden may trigger. If this is right, then the familiar image of the garden "mediating" between human creativity and the natural order needs to make proper room for nature's fourth dimension, and not

14 *What Gardens Mean*, 49–84, 85–120.

15 For a comparison of gardens with music see: Barwell and Powell, "Gardens, Music, and Time," 142–7. For comparisons of gardens with music, dance, kinetic sculpture and environmental art see: Powell, "We Do Not Have an Adequate Conception of Art until We Have One That Accommodates Gardens," 86–94. For comparisons with drama and other performing art see: Salwa, "The Garden as a Performance," 373–83.

16 Miller calls gardens *Gesamtkunstwerke*. (Mara Miller, "Gardens: Gardens as Art," in *Encyclopedia of Aesthetics*, ed. Michael Kelly (New York, Oxford: Oxford University Press, 1998), 274.)

> just its spatial aspects. An interesting garden may "borrow" the weather's impending change as much as the distant mountain scenery.[17]

How a visitor experiences a garden on a first or subsequent visit is necessarily influenced by the facts that the visits themselves are temporal events and gardens are temporal objects. I believe that the implications of both facts have been underplayed, certainly in historical but also in some recent philosophical garden writing. Straightforward questions that gardens' temporality invite include: why have the roses (not) been dead-headed? Is this garden finished yet? What time of the day/year/month is the best time for visitors to see the garden? Why are the blue poppies not flowering today? Straightforward questions that the experience of gardens invite include: which direction(s) should I walk in along this path? And, how long does it take to see this garden? Questions such as these, and their answers, are taken for granted by gardeners and garden visitors, but the questions, answers, and their implications, have only recently attracted any serious philosophical interest. Furthermore, an encounter with a garden is not merely one single durational event. Within the overall duration of a visit, a visitor is constantly making temporal decisions at a shorter scale, at the same time as she is making locational decisions. To paraphrase Birksted words quoted earlier, a garden beholder is perpetually and freely moving in four dimensions.

Gardens' art status, and the related artist status of garden makers, have presented some problems for contemporary philosophical aesthetics. These problems may be understood to be ones which, to varying degrees, frustrate the application of the term "art" and "artist" to gardens and garden makers. These are not new matters for gardens and garden makers to deal with. They are problems that have persisted, albeit for differing reasons, for 200 or so years. However, my concern is not with historical but with contemporary definitions of art, by which I mean the relational-historical-institutional definitions of art which have emerged over the last 50 years. I am interested in them because it is in the context of such definitions that gardens have, until most recently, remained outside the category of art.

17 Cooper, "Foreword," ix.

That this has happened is puzzling because I believe that representative examples of these definitions, such as those developed by the contemporary philosophers George Dickie, Jerrold Levinson, Noël Carroll, and the late Arthur Danto, allow in theory for garden-like objects to be works of art. In spite of this, the definitions of Danto and Dickie have been used by at least two philosophers to support their claims that gardens are not art.[18]

The fact that these recent definitions *can* admit gardens to the category of art but have, with the two negative exceptions noted above, not been applied in this way, suggests that the question of gardens-as-art is not so much a matter of theoretical difficulty as a matter of uninterest for philosophers of art. From the philosophers' point of view, there are plausible reasons for their uninterest. After all, gardens tend to be attractive and direct rather that confrontational and ironic, and pleasingly sensual rather than densely intellectual. In fact, understood in these terms, most gardens appear somewhat quaint and old-fashioned, not at all the sort of art objects contemporary aestheticians should be troubled by. To be plain, most gardens are not fancy-pants art.[19]

However, this apparent uninterest in gardens has been to philosophy's disadvantage, and not only because it has resulted in philosophers' misunderstanding and neglecting gardens themselves. It turns out that the very qualities that gardens have always possessed, such as instability, lack of distinct spatial and temporal boundaries, multiple authorship, multi-sensuality, and ordinary "objectness," are all hallmarks of much of the "new" avant-garde, or should that be avant-garden, arts of the twentieth and twenty-first centuries. An adequate understanding of gardens can, therefore, be a useful tool for advancing the understanding of some new, challenging art forms, such as computer and street art, and it is disappointing that philosophy has not made use of this tool.

The status of, say, Mozart as composer-artist of *The Marriage of Figaro*, and Mies van der Rohe as architect-artist of the Barcelona Pavilion, are well established. Equally, the roles of their respective collaborators and

18 For Miller's invocation of Dickie's definition and Ross's of Danto's see: Miller, *The Garden as an Art*, 69–71; Ross, *What Gardens Mean*, 189–208.

19 I am indebted to Dr Sondra Bacharach for this apt descriptor.

contributors, such as musicians, librettist, builders, stone-masons, and so on, are acknowledged, without thereby undermining Mozart's and Mies van der Rohe's status as artists. However, for three reasons, the case of the garden-artist is more complex and deserving of philosophical attention. First, the installation of gardens does not always – even usually – result from a fully detailed set of drawings or other instructions produced by a designer-artist and covering all aspects of the project. Much, unspecified detail is left to the expertise and experience of builders and gardeners. This is similar to the case of, say, mounting an operatic production, except that in the case of opera, the principal musical details are well-specified and deviations from them are only tolerated within fairly strict, commonly agreed limits. Second, unlike the first performance of a play, or the completion of a building, painting, or poem, the initial installation of a garden almost always bears only slight resemblance to its "final" mature instance.[20] In a garden, decisions are made and interventions are carried out continually, sometimes over centuries, by those other than the garden's originator, and these can have profound effects on the appearance, and our experiences of a garden. Third, left to its own devices, the Barcelona Pavilion would have deteriorated but still have remained, for a long time, recognizable as the same building.[21] The case of gardens is different. Left to their own devices, many, even most, would become unrecognizable in a relatively short time. Disease, death, weeds, and climate would all ensure that. The ongoing contributions of managers, owners, gardeners, horticulturalists, pedologists, and so on, are essential to the survival of a garden if it is to remain similar to that envisaged by the original garden maker-artist. Because of this, gardens, garden-artists, and garden contributors and collaborators occupy a niche which, until recently, they have had to themselves. However, new developments in some arts, including architecture, and the appearance of new artforms, including environment or land art, computer, and street art mean that these "problematic" characteristics of garden art and artistship

20 There is, of course, no such thing as a final version of a garden.

21 Interestingly, the Barcelona Pavilion was pulled down and then reconstructed 56 years later. It looked (almost) exactly the same when reconstructed as it had originally. In the case of gardens, this is not a possibility.

are now being addressed by philosophy, albeit in the context of these new, emerging artforms, rather than in gardens.

In summary, gardens have been, and for some philosopher-aestheticians continue to be, problematic members of the art category. Stumbling blocks to full membership of the category, as traditionally construed, include gardens' ontology, their typically wide sensory range, their unique temporal-experiential qualities, and the difficulties they have posed for definitions of "art" and for the application of the term "artist." Examining the historical context which has allowed this state of affairs to develop is the principal task of Part II.

PART II

The Territory: A Survey

CHAPTER 2

Changing Views: Philosophical Aesthetics (1700–2017)

Over the last 300 or so years, the relationship between gardens and philosophical aesthetics has varied, especially when it is viewed in the context of the wider philosophical aesthetic endeavours of the period. During this time, the "art" status of gardens has been influenced as much by what was going on "beyond the pale" as it was by changes occurring in gardens themselves. The period encompasses the beginning of modern aesthetics, the rapid development of aesthetics during the eighteenth century, aesthetics' relative quiescence during the nineteenth century, and its new flourishing in the twentieth and twenty-first centuries. During this time, aesthetics has directed its attention to different paradigmatic object categories including nature and the natural environment, the human environment, everyday objects and activities, and art.

The term "aesthetics" was coined in its modern usage by the philosopher Alexander Baumgarten in the early eighteenth century.[1] Baumgarten employed the term to mean the ability to not only experience but also to judge sensory data, particularly with regard to whether or not the object or event being perceived exhibited beauty. This exercise of judgement he called "taste." It was a judgement made via the senses, not via the intellect. Put very simply, a judgement of beauty with respect to an object or event was understood to be based on whether the experience of that object or event gave rise to sensory pleasure of the appropriate kind in the perceiver. Furthermore, it was hoped that, based on a careful assessment of individual

1 Darren Hudson Hick, *Introducing Aesthetics and the Philosophy of Art* (London and New York: Continuum, 2012), 1.

experiences of taste, principles, or rules of what constituted beauty might be able to be generated.[2]

During the eighteenth century, inquiry into the theory and practice of aesthetics flourished. Technical philosophical arguments concerning the details and nature of the mechanism(s) by which judgements of taste were made proliferated. Were such judgements purely sensory, were they intellectual, or imaginative, or did they involve some combination of these faculties? For example, Joseph Addison argued that aesthetic judgements involved the imagination, the Earl of Shaftesbury that they were importantly mental processes, and Kant that they involved the free play of both the understanding and the imagination.

The eighteenth century was also witness to an expansion in the range of sensory qualities deemed to be of aesthetic interest and importance. As the century progressed, the sensory experience of beauty described by Baumgarten was joined by sensory experiences of the sublime and the picturesque as matters fit for aesthetic experience.

At this time, varying opinions were held regarding what object, or objects, were the paradigmatic objects of aesthetic attention. For Baumgarten, nature and art were equally plausible candidates, and most eighteenth-century aestheticians thought similarly. However, by the end of the century, the publication of Kant's influential *Critique of Pure Judgement* in 1790 ensured that his view, that nature was the preeminent and paradigmatic aesthetic object, appeared as the authoritative, if retrospective, last word on the matter for the century.[3]

In his *Critique of Pure Judgement*, Kant distinguishes between experiences of the beautiful and the sublime on one hand and the merely agreeable on the other. The *Critique* also distinguishes between experiences of what Kant called pure beauty and adherent beauty. It is not necessary to elaborate here on these important distinctions other than to note the important role

2 For a summary account of "taste" in the context of eighteenth-century Britain see: James Shelley, "18th-Century British Aesthetics," ed. Edward N. Zalta, *The Stanford Encyclopedia of Philosophy* (Fall 2014 Edition), <http://plato.stanford.edu/archives/fall2014/entries/aesthetics-18th-british/>. Introduction.

3 Kant, *Critique of Judgement*.

the concept of "disinterestedness" plays in establishing and underpinning them.[4] That concept has remained a fundamental if contested notion of the philosophy and, at least in theory, the experience of art, and of nature, from the eighteenth century until the present day.

Put simply, disinterestedness characterizes a preferred mode of aesthetic attention to an object or event. It involves paying attention to that object or event without any thoughts concerning the object's or event's usefulness, desirability, value, or fitness for function. While such detachment may be comparatively easily achieved by someone experiencing a beautiful natural event, such as a sunset, the desired detachment appears more difficult to achieve when experiencing a beautiful painting, teapot, or person. The concept of "disinterestedness" has played, and, for some, continues to play, an important role in how we appreciate nature, the human environment, and art.

Kant's writings effectively brought eighteenth-century philosophical aesthetics to a magisterial close. They continue to be influential and much-debated to this day. However, Kant's status as the preeminent aesthetician of his time was not long lived. In 1818, G. W. F. Hegel delivered his series of lectures on aesthetics for the first of many times.[5] Kant and Hegel differed with respect to their accounts of the nature of aesthetic judgements and also, and more importantly for this study, with respect to what each regarded as the paradigmatic object of beauty. Kant had proposed nature, but Hegel now proposed art. This chance of focus was highly influential; so much so that for the following 130 or so years, aesthetics came to be equated almost exclusively with the sensory perception of beauty and other qualities in works of art, not in nature.[6]

4 Ibid., § 6, § 7.

5 Stephen Houlgate, "Hegel's Aesthetics," ed. Edward N. Zalta, *The Stanford Encyclopedia of Philosophy* (Spring 2014 Edition), <http://plato.stanford.edu/archives/spr2014/entries/hegel-aesthetics/>. § 2.

6 For accounts of the Hegel's preference for art over nature as the paradigmatic art object, see: Douglas Burnham, "Immanuel Kant: Aesthetics," ed. James Fieser, *The Internet Encyclopedia of Philosophy* (2014), <http://www.iep.utm.edu/>. § 2d; Allen Carlson, "Environmental Aesthetics," ed. Edward N. Zalta, Summer 2012 edn, *The*

After the heyday of philosophical aesthetics, which stretched for 100 or so years from the time of Baumgarten and culminated in the work of Kant and Hegel, philosophical aesthetics went into a period of relative inactivity, from which it was only roused by the upheavals that began to occur in the world of art in the late nineteenth century and that continued into the first half of the twentieth century. When philosophical aesthetics re-emerged at that time, its focus remained firmly on art, as opposed to nature, but its concerns with art were different to those which had occupied aestheticians a century earlier. In particular, aesthetics became increasingly concerned with providing a framework adequate to understanding the new abstract and, later, the seemingly non-artworks being produced by visual and other artists.

Around the middle of the twentieth century, when the philosophy of art was grappling with its "anxious objects" and moving towards an aesthetics of art based on non-traditional, *extrinsic* properties of artworks, new schools of aesthetics were emerging which, at least initially, maintained a more traditional interest in the *manifest*, or *intrinsic*, properties of objects and events.[7] There were three of these new schools: the first revisited, albeit with very different conceptions and assumptions, the eighteenth-century interest of aesthetics in nature and our experiences of it, and in time this school became part of the larger discipline of the philosophy of the environment; the second school was interested in the aesthetics of the human (or built) environment and shared this interest with the philosophy of architecture and, by extension, in part with the philosophy of art; the third school broke completely new ground by taking as its focuses of interest everyday objects, practices, and events.[8]

Stanford Encyclopedia of Philosophy (Summer 2012), <http://plato.stanford.edu/archives/spr2015/entries/environmental-aesthetics/>. § 1.2.

7 The useful term, "anxious object," was introduced in: Harold Rosenberg, *The Anxious Object: Art Today and Its Audience* (New York: Horizon Press, 1964).

8 For overviews of the evolution and current state of these disciplines, see: E. Brady, *Aesthetics of the Natural Environment* (Tuscaloosa: University of Alabama Press, 2003); A. Carlson and A. Berleant, eds, *The Aesthetics of Human Environments* (Peterborough, Ontario: Broadview Press, 2007); Thomas Leddy, *The Extraordinary*

To different degrees and with differing emphases, these three new schools of aesthetics all contested the concept of "disinterestedness," or "psychical distance," which had long been a foundational concept in art aesthetics. They also contested the primacy that traditional aesthetics had accorded the senses of sight and hearing, and they brought taste, touch, smell, and kinaesthesia into the fold of aesthetically acceptable senses.[9] The challenging of these concepts mirrored similar concerns beginning to occur during the latter part of the twentieth century in the philosophy of art and had important implications for what came to be counted as an adequate, or valid, experience of art.

Aesthetics of Nature

Plato and Aristotle had strongly contrasting views on the value of the representational, or mimetic, activity that came to typify art through until the late nineteenth century, but they were agreed that representation, and in particular the representation of what they referred to as nature, was the essential activity of art.[10] For them, what was represented in art was not,

---

in the Ordinary: The Aesthetics of Everyday Life (Peterborough, Ontario: Broadview Press, 2012).

9 Note, however, that immersive, multi-sensory aesthetics had not been unknown to philosophers of art within the pragmatist tradition. A pioneering exposition of it was published in 1934 by the American pragmatist John Dewey. See: J. Dewey, *Art as Experience* (New York: Putnam Capricorn, 1958). For contemporary expositions of differing versions of immersive, multi-sensory aesthetics from outside the mainstream analytic tradition, see, for example, the work of the pragmatist-influenced American philosopher Arnold Berleant and the work of the American neo-pragmatist Richard Shusterman: Arnold Berleant, *Art and Engagement* (Philadelphia, PA: Temple University Press, 1991); Richard Shusterman, *Body Consciousness: A Philosophy of Mindfulness and Somaesthetics* (New York: Cambridge University Press, 2008).

10 For Plato, art (merely) represents nature and therefore it "ranks far below the truth. … [It] corrupts the soul," and ought to be banned from well-functioning cities. (See:

say, the person depicted in a painting, nor the real or imagined person who served as the model for the painted depiction of the person. Rather, what was represented was an originary or *ur*-person – in other words "personness," or human nature.[11] And the role of this representation was not merely aesthetic. It was also in part didactic. For Aristotle, such representation concerned human nature as it *ought* to be, "raised ... above all that is local and accidental, so as to be in the highest sense representative."[12]

This idea, that mimesis in art is a representation of some originary object, quality, form, or designing mind rather than of the mere object or representing object available to the immediate senses, was enormously influential through to and beyond the eighteenth century. When eighteenth-century aestheticians and philosophers talked of art imitating nature, they were still referring to what from now on I will distinguish from its everyday use by using an upper-case initial "N," Nature. For them, Nature did not refer merely to their surroundings of plants, animals, mountains, and so on. For them, imitated Nature encompassed not only "natural" nature but also human nature, divine nature, and the nature of the universe.[13] And this was, in part, what the English poet Alexander Pope had in mind when, in his *An Essay on Criticism* (1710), he famously advised would-be artists "to copy Nature."[14]

Plato, "The Republic: Book X," 596e-608b.) For Aristotle, mimesis represents nature and is therefore to be valued because it enlightens us by depiction, from which we can learn about the world, and how better to behave. (See: Aristotle, "Poetics," 1148b 4–24.)

11 See: Plato, "The Republic: Book X," 595a-608b.

12 Quoted in: Rensselaer W. Lee, *Ut Pictora Poesis: The Humanistic Study of Painting* (New York: Norton, 1967), 9.

13 For varying accounts of this transition from "Nature" to "nature" see: Steven Heyde, "The Historical Roots of 'Aesthetics' in Landscape Architecture: An Introduction," *Studies in the History of Gardens & Designed Landscapes* 34, no. 2 (2014), 124; Tom Turner, *Garden History: Philosophy and Design 2000 B.C.–2000 A.D.* (London and New York: Spon Press, 2005), 226; Raymond Williams, *Keywords: A Vocabulary of Culture and Society* (New York: Oxford University Press, 1976), 219.

14 Alexander Pope, *An Essay on Criticism*, <https://www.gutenberg.org/files/7409/7409–h/7409-h.htm>. Part I, 133–40.

This received conception of Nature was still largely in place around the beginning of the eighteenth century. However, that same time marked the start of the decline and final disappearance of it and, simultaneously, the rise of the acceptability of nature *tout court* as something worthy of artistic representation and, most importantly for aesthetics, as something that potentially exhibits beauty. The philosopher Glenn Parsons offers some interesting reasons why this change in nature's status was unable to happen earlier. He claims that earlier views of nature were based on mathematical harmonies that were hard to find in nature, especially in a world where microscopes were rare; that there were few opportunities for most people to experience nature before 1800; and that religious beliefs, such as the belief that mountains were unattractive debris left behind after the Biblical flood, mitigated against seeing beauty in nature.[15] Parsons also offers reasons why nature did indeed became an acceptable aesthetic object during the eighteenth century. His reasons include the emergence of a new appreciation of the Alps, stemming from the resumption of the "Grand Tour" after the conference of Utrecht; the influence of the paintings of Rosa, Lorraine, and Dughet, also as a result of the resumption of the Grand Tour; and developments in geology, and science's conception of mountains as natural phenomena rather that Biblical detritus.[16] Furthermore, there exists ample evidence from contemporary philosophy, for example in the writings of Kant and the philosopher-botanist Jean-Jacques Rousseau, and from contemporary art, for example in the poems of Wordsworth and the paintings of Constable, that such was the case.[17]

One of nature's most significant roles during the century was as exemplar of the important eighteenth-century aesthetic qualities: the beautiful, the sublime, and the picturesque. Environmental philosopher Allen Carlson

15 Parsons, *Bloomsbury Aesthetics: Aesthetics and Nature*, 7–8.

16 Ibid., 8–9.

17 See Christopher Bertram, "Jean Jacques Rousseau," ed. E. N. Zalta, *The Stanford Encyclopedia of Philosophy* (Winter 2012), <http://plato.stanford.edu/archives/win2012/entries/rousseau/>. Introduction. I discuss Kant's privileging of natural beauty over art as the paradigmatic aesthetic object at 38–9.

usefully summarizes the differences between these three aesthetic qualities as they appeared to the eighteenth-century mind:

> Objects experienced as beautiful tend to be small and smooth, but subtly varied, delicate and "fair" in colour, while those experienced as sublime, by contrast, are powerful, vast, intense, terrifying, and "definitionless." Picturesque items are typically in the middle ground between those experienced as either sublime or beautiful, being complex and eccentric, varied and irregular, rich and forceful, and vibrant with energy.[18]

Carlson goes on to say, "the idea of the picturesque, rather [than] that of the beautiful or the sublime, achieved the greatest prominence concerning the aesthetic experience of nature."[19] This was the case, he argues, because, to the eighteenth-century mind, nature more readily and obviously reflected the picturesque characteristics outlined above. He also suggests that the eighteenth-century's emphasis on the importance of disinterestedness in aesthetic experience required that nature be stripped of any personal associations, pleasures, or uses it may have had for a viewer and that this in turn paved the way for landscapes being appreciated picturesquely, that is, as pictures.

Although Kant wrote extensively on beauty and the sublime, he did not consider the picturesque, the promotion of which fell to the Englishmen William Gilpin, Uvedale Price, and Richard Payne Knight. Popularized by the later eighteenth-century writings of these men, which ranged from travel guides to philosophical essays and poetry, the picturesque became the dominant mode of the aesthetic appreciation of nature for the next 150 or so years.[20]

18 Allen Carlson, "Environmental Aesthetics," ed. Edward N. Zalta, Summer 2012 edn. Ibid. (Summer 2012), <http://plato.stanford.edu/archives/spr2015/entries/environmental-aesthetics/>. § 1.1.

19 Ibid.

20 See as examples: William Gilpin, *Three Essays: On Picturesque Beauty; on Picturesque Travel; and on Landscape Sketching: To Which Is Added a Poem on Landscape Painting* [electronic resource] (London: R. Blamire, 1792); Uvedale Price, *Essays on the Picturesque, as Compared with the Sublime and the Beautiful; and, on the Use of Studying Pictures, for the Purpose of Improving Real Landscape*, 2 vols [electronic

Once nature was displaced by art as the paradigm object for aesthetic attention, the philosophical aesthetics of nature remained largely unaddressed during the remainder of the nineteenth century and, in fact, until the middle of the twentieth century. However, although nature was ignored by philosophers of aesthetics, it was eagerly taken up as a subject appropriate to art by painters, composers, poets, and novelists. Nature, in the form of landscape, was still co-opted for its traditional aesthetic values, particularly for its sublimity, but landscape was equally employed as an expressive, symbolic, or metaphorical participant in and setting for human activities. According to the art historian Kenneth Clark, landscape painting was "the chief artistic creation of the nineteenth century."[21] Landscape also flourished as subject matter for other art forms also: consider, for example, the stream in Schubert's *Die schöne Müllerin*, the daffodils in Wordsworth's eponymous poem, the heath named Egdon in Hardy's *The Return of the Native* and, in the twentieth century, the waterfall in Frank Lloyd Wright's Fallingwater house.[22] In each of these examples, nature is more than a mere backdrop: it has become an important participant in the work, a *dramatis persona* in its own right, as well as, in some cases, an amplifying reflector of the moods and emotions of the works' human protagonists. It is not hard to foresee the progression from here to the situation which the philosopher Ronald Hepburn describes with respect to the twentieth century, when

resource] (London: J. Mawman, 1810); Richard Payne Knight, *The Landscape: A Didactic Poem in Three Books: Addressed to Uvedale Price, Esq*. [electronic resource] (London, 1795).

21 Kenneth Clark, *Landscape into Art* (London: J. Murray, 1976), 15.

22 However, there was a strand of romanticism which, while still revelling in the natural world, believed that happiness was to be found anywhere other than the (natural) place in which one found oneself. This strand reached one of its apogees in von Lübeck/Schubert's famous lied, *Das Wanderer*, which reaches it climax with the line, "*Dort, wo du nicht bist, ist das Glück*" [literally: "There, where you are not, is happiness"]. Such an attitude *may* have contributed in part to the nineteenth-century's diregard for gardens, which are, if nothing else, actual, natural places where one might expect to be happy. I discuss gardens in the context of romanticism at 62–6.

artists came to prefer the "inner landscape of the human psyche" to that of the natural world.[23]

At this time of transition between Enlightenment and romantic values and aesthetics, the meaning of "nature" moved from referring to the world of universal forms to what Tom Turner has called "the world of particulars." It was, he says, "a change from the '*nature* of the world' to the 'world of nature.'"[24] Although he is describing a change of focus that occurred in the world of aesthetics, that change was reflective of bigger changes occurring in the worlds of philosophy and, increasingly, science. This transition from the Enlightenment to the romantic Era was marked not only by a shift from aesthetic neo-classicism to romanticism: equally important at that time was the increasing dominance of empiricism over rationalism in the fields of philosophy and the natural sciences.

According to Carlson, the increasing importance of the empirical natural sciences from this time was a vital component in the early development of what grew into one of the dominant strands of late twentieth-, and twenty-first-century philosophy: the philosophy of the environment.[25] He claims that contemporary developments in the natural sciences, including geography, were increasingly an influence on the "nature writing" of Thoreau and others that was emerging in North America in the first half of the nineteenth century. This nature writing began with a picturesque point of view but ended up representing a new way of appreciating nature, exemplified for Carlson in the writing of the naturalist John Muir. In his writings, Muir explicitly rejects the validity of a picturesque appreciation of nature. To him, all nature was beautiful, whether it looked like a picture or not. Near the end of the nineteenth century he wrote, "God never made an ugly landscape. All that the sun shines on is beautiful, so long

23 R. W. Hepburn, "Contemporary Aesthetics and the Neglect of Natural Beauty," in *British Analytical Philosophy*, eds Bernard Williams and Alan Montefiore (London: Routledge and Kegan Paul, 1966), 287.

24 Turner, *Garden History: Philosophy and Design 2000 B.C.–2000 A.D.*, 190.

25 Carlson, "Environmental Aesthetics," § 1.2.

as it is wild."[26] Carlson calls this point of view "positive aesthetics," and describes it as "the converse of aesthetic appreciation influenced by the picturesque, which finds delight in evidence of human presence."[27] Muir therefore emerges as an important precursor of the environmental movement and as a pivotal figure in the development of the philosophy of the environment in the twentieth century.

During the first half of the twentieth century, philosophical aesthetics continued largely to ignore the aesthetics of nature. Aestheticians were preoccupied by the exciting, dynamic, problematic nature of contemporary art and they had no inclination to pay regard to "unproblematic" nature, whose "problems" appeared to have been sorted out already. Those who did pay any attention to nature did so through the filter of the aesthetics of art. However, the inclination to ignore nature, or to treat it *qua* art, began to be reversed from the 1960s onwards. Contributing to this reversal was the birth of the environmental movement, whose concern was to halt, limit, or even reverse damage to the natural world, and, in the world of philosophical aesthetics, the publication of Hepburn's landmark 1966 paper, "Contemporary Aesthetics and the Neglect of Natural Beauty."[28]

Hepburn's paper may uncontroversially be seen as representing the turning point when philosophical aesthetics began to have a renewed and invigorated regard for nature – a trend which continues apace today.[29] My interest in Hepburn's paper acknowledges its importance for environmental philosophy, but also acknowledges its importance for the philosophy of art and, especially, for the philosophy of gardens and some contemporary art forms, including installations, and land art.

26 John Muir, "The Wild Parks and Forest Reservations of the West," *The Atlantic Monthly* 81, no. 483 (1898), <http://vault.sierraclub.org/john_muir_exhibit/writings/favorite_quotations.aspx>.

27 Carlson, "Environmental Aesthetics," § 1.2.

28 Hepburn, "Contemporary Aesthetics and the Neglect of Natural Beauty."

29 Parsons dates analytic philosophy's renewed interest in nature and its new interest in the environment from the publication of Hepburn's paper: G. Parsons, "The Aesthetics of Nature," *Philosophy Compass* 2, no. 3 (2007), 358. And another important philosopher of the environment, Allen Carlson, describes Hepburn's paper as seminal: Carlson, "Environmental Aesthetics," § 2.2.

Hepburn offers two principal reasons why what he considers an appropriate appreciation of nature ought to differ from the earlier, aesthetic approach to nature appreciation. First, nature surrounds and engulfs us. We are in it and it is in us. We move in it. We depend on it and we cannot easily be disinterested, traditionally aesthetic observers of it. Second, nature is frameless. *We* draw the boundaries of our perceptual experiences of nature. It is not formally complete. We set the contexts and limits for our spatial experiences of it. It is therefore inappropriate to experience nature as a standalone, self-sufficient, clearly defined, static, traditional aesthetic object.

Hepburn elaborates on the notion of contextualism in a way that is interesting. He extends his concept of spatial expansion of context to what might be called a temporal expansion of context. In doing this. he importantly acknowledges that natural objects can equally be seen as natural *processes*. For example, Hepburn writes of walking across an expanse of sand and mud with the knowledge that he is walking in a tidal basin at low tide. That knowledge, he writes, "is not aesthetically irrelevant. I see myself now as … walking on what is for half the day sea-bed. The wild emptiness may be tempered by a disturbing weirdness."[30]

In sum, Hepburn claims that our experiences of nature are immersive and spatially and temporally open-ended. Our experiences of nature are also always provisional, and depend on the contexts and focuses that we, as participants, set up. These claims have been of considerable significance in the development of environmental aesthetics and nor have they been without influence in the developments of the new fields of the aesthetics of the human environment and the everyday.

30 Hepburn, "Contemporary Aesthetics and the Neglect of Natural Beauty," 295.

Aesthetics of the Human Environment and the Everyday

Once the aesthetic appreciation of nature was divorced from a foundational reliance on picturesque aesthetics and new, more appropriate ways of describing the natural environment and our interactions with it were being established, the time was ripe for the new aesthetics to be extended to fields not previously considered worthy of aesthetic attention, namely the human, or built, environment, and everyday objects and activities. The aesthetics of the human environment involves settings as diverse as farms, shopping malls, sports stadiums, funeral parlours, and airports and everyday aesthetics embraces activities as different as wine tasting, dusting, sex, sport, and cosmetic dentistry.[31] Just as was the case for the revolutionary nature aesthetics that enabled it, this extension of aesthetics' fields of interest had important implications for the aesthetics of art in general and gardens in particular.

One of the leading figures in the aesthetics of the human environment and the everyday is the philosopher and composer Arnold Berleant. Principal among his claims is that disinterestedness has never been an adequate way of characterizing the aesthetic attitude, whether that attitude be directed towards art, nature, or anything else. Instead he argues for what appears, at least superficially, to be the exact opposite of disinterestedness: total engagement. He characterizes aesthetic engagement in this way: "Aesthetic engagement emphasizes the holistic, contextual character of aesthetic appreciation. Aesthetic engagement involves active participation in the appreciative process, sometimes by overt physical action but always by creative perceptual involvement."[32] Furthermore, aesthetic

31 There is a rapidly increasing amount of material available in these fields. Good overviews include: A. Berleant and A. Carlson, "Introduction," in *The Aesthetics of Human Environments*, eds A. Berleant and A. Carlson (Peterborough, Ontario: Broadview Press, 2007); Leddy, *The Extraordinary in the Ordinary: The Aesthetics of Everyday Life*; Yuriko Saito, *Everyday Aesthetics* (New York: Oxford University Press, 2008).

32 Arnold Berleant, "What Is Aesthetic Engagement," *Contemporary Aesthetics* 11 (2013), <http://www.contempaesthetics.org/newvolume/pages/article.php?articleID=684>.

engagement ensures that "perception itself is reconfigured to recognize the mutual activity of *all* the sense modalities, including kinaesthetic and somatic sensibility more generally" (my emphasis).[33] Berleant's concept of aesthetic engagement is useful for understanding our experiences of the human environment and everyday activities, and also, potentially, our experiences of gardens.[34]

Aesthetics of Art

According to the philosopher Oskar Kristeller, it was the Abbé Batteux who, in his famous treatise of 1746, *Les beaux arts réduits à un même principe*, finally gave clear and coherent expression to the formulation we still take for granted as the basis of the modern system of the fine arts.[35] Batteux's fine arts included music, poetry, painting, sculpture, and dance. He claimed that the unifying factor, the *raison d'être*, shared by these disparate activities was their purpose in affording an "imitation of beautiful nature."[36] In making this claim, Batteux set important parameters of inquiry for the aesthetics of art. The questions that he sought to answer, namely "What is art?" and "What does art do?", are among the questions that have been repeatedly asked and variously answered by aestheticians right to the present day. However, while aestheticians have continued to ask similar question

33 "What Is Aesthetic Engagement?," *Contemporary Aesthetics* 11 (2013).

34 However, a detailed engagement with Berleant's pragmatically and phenonmenologically influenced aesthetics is beyond the scope of this study.

35 Oskar Kristeller, "The Modern System of the Arts," in *Renaissance Thought 2; Papers on Humanism and the Arts* (New York: Harper and Row, 1965), 163–227. For a part translation of Batteux's text see: Susan L. Feagin and Patrick Maynard, eds, *Aesthetics*, Oxford Readers (Oxford: Oxford University Press, 1997), 102–4.

36 *Aesthetics*, 103.

of art, the contextual settings of their questionings, and therefore their answers, have varied significantly.[37]

During the early Enlightenment, aesthetics as formulated by Baumgarten applied equally to works of art and nature. Answers to any aesthetic question regarding art could be adequately answered, *mutatis mutandis*, with reference to either art or nature, and, almost exclusively, aesthetics related to data perceived through the sense of sight only. However, Kant challenged those views. At the time of the publication of his third *Critique* in 1790, Kant came down firmly on the side of nature as the paradigmatic aesthetic object.[38] For him, art was a secondary concept which borrowed its qualities of beauty or sublimity from nature. He argued that sight was not necessarily the preferred aesthetic sense organ and considered hearing as a potentially valuable receiver of aesthetic sensory data.[39] For Kant, music was something capable of beauty but something that was, in the end, trivial when compared with the other fine arts. Kant also extended his regard to the sense of smell, and the philosopher Marta Tafalla has recently interrogated his account of smell in the context of gardens.[40]

According to Hepburn, the eighteenth-century aesthetic treatment of art had been secondary and derivative.[41] However, this situation changed shortly after Kant's death, when his role as the leading philosophical aesthetician of the day was assumed by Hegel. Hegel was a sophisticated, artistically literate man, and his work contains important assessments of most of the fine arts as well as accounts of individual works of art in the Western tradition from ancient times onwards. The delivery of his lectures on art, and their later publication by his students, mark an important turning

37 For a recent, interesting critique of Batteux's unified system of the arts see: John Macarthur, "Architecture and the System of the Arts; or, Kant on Landscape Gardening," in *Architecture, Disciplinarity, and the Arts*, eds Andew Leach and John Macarthur (Ghent: A. & S. Books, 2009).

38 See: Kant, *Critique of Judgement*, § 45.

39 See: ibid., § 53. However, it would be unwise to infer from this that Kant was a serious music lover. His musical taste did not extend beyond marching music, which is surprising for one whose contemporary countrymen included Haydn and Mozart.

40 Tafalla, "Smell and Anosmia in the Appreciation of Gardens."

41 Hepburn, "Contemporary Aesthetics and the Neglect of Natural Beauty," 288.

point in aesthetics. Their publication coincides (roughly) with the transition between (neo)classicism and romanticism, which it reinforces, the emergence of the individual as an important element in society and art, and the emergence of new political orders.[42] At this time, the (neo)classicism of the eighteenth century gave way to an impatient, emotional, individualized romanticism, and Hegel's claim that art represented the highest expression of an individual artist's and a culture's spirit was well suited to its time and has remained enormously influential into the twenty-first century.

Hegel's work also marks a turning point between the representative, or mimetic, conception of art, which had been dominant since Ancient Greece, and the new expressive conception of art which replaced it. The expressive conception of art that Hegel and other nineteenth-century philosophers subscribed to stressed the interrelationships between art and human emotions, aspirations, and spirituality, and it remained dominant throughout the romantic era. However, around the turn of the twentieth century, Hegel's view of art was being supplanted by formalist conceptions. This was the period of "art for art's sake." Art was to be understood, valued, and experienced as standalone, "formal," aesthetic objects or events, without reference to any narrative or representative qualities those objects or events might possess.[43] And art was by now the sole interest of philosophical aesthetics.

The formalist Clive Bell's *Art* appeared early in the twentieth century.[44] Bell was a champion of contemporary impressionist and post-impressionist art and his conception of art as significant form was developed, at least in part, to facilitate the comprehension of the increasingly abstract visual art being produced by painters belonging to these schools. However, the notorious artworks which subsequently began to appear dealt body blows to

42 Houlgate, "Hegel's Aesthetics," § 2.

43 However, the nineteenth-century conception of art as expression did not disappear. It continued to be promoted well into the twentieth century by Croce and Collingwood, for whom it was a fundamental component of their conception of art as mental product and activity. But, by this time, the scope of "expression" had been expanded to include much more that just the expression of an artist's felt emotions.

44 Clive Bell, "Significant Form (Extracts from Chapter One and Chapter Three of *Art*)," in *Aesthetics: Twentieth Century Readings*, ed. E. Hirst (Boston, MA: Beacon Books, 1986).

eighteenth-, nineteenth-, and early twentieth-century notions that equated art with beauty, significant form, or the expression of an artist's inner world. These new works called into question the very category of "art" itself.

In the process of trying to comprehend these and similar avant-garde works, the aesthetics of art transformed itself into the philosophy of art, the focus of which was increasingly on the ways in which a work exists, and relates to its society and the history of art. Aesthetics, or the philosophy of art, came to concern itself with ontological, relational, and extrinsic properties of a work of art rather than with a work's manifest properties, which had previously been understood to be the appropriate subject of aesthetic investigation.[45]

The philosophy of art became increasingly focused on questions, the answers to which could, up until that time, have been taken for granted. For example: what is a work of art? Can a work of art be created by chance? Is there a single definition of art? Can music be silent? How can we identify a work of art? How do works of art exist? Are disgust and ugliness valid aesthetic properties? And, it was at about this time that the term "aesthetics" was taken up by other philosophers and applied to the fields of the environment and the everyday.

Gardens, and Conceptions of Art

I have briefly outlined four conceptions of art above. Put simply, mimetic art represents something, expressive art expresses something, formalist art exhibits form, and, according to the relational theorists, individual works of art relate to culture and society in particular ways. Examining gardens in the contexts of different conceptions of art is not a focus of this study.[46] However, art gardens, just as much as, say, poems, paintings, and symphonies, have always had the potential to be fully "contemporary," by which

45 See, however, footnote 14, p. 8.

46 For an overview of that topic see: Powell, "We Do Not Have an Adequate Conception of Art until We Have One That Accommodates Gardens."

I mean expressive of the social and intellectual milieu in which they were first formed. Each of the four conceptions of art can readily be paired with paradigmatic art gardens which express well their particular values and emphases (see Figures 13–16).

Mimesis was the dominant conception of art in the Western world for by far the greatest part of Western art's history. On this conception, eighteenth-century mimetic art represented not simply what it depicted in the first instance. It also represented – or perhaps re-presented is a more apposite term – objects and events at increasingly distant metaphysical levels from the actual physical objects and events with which a viewer was engaging. Appropriate engagement with this art entailed aesthetic, intellectual, and didactic experiences. A garden typical of this conception is the famous garden at Stowe in England, where garden elements represent a range of political, spiritual, and artistic concepts (see Figure 13).

The expression theory replaced mimesis as the predominant de facto conception of art as the nineteenth century progressed. According to this theory, the essence of art was concerned with the expression of an artist's

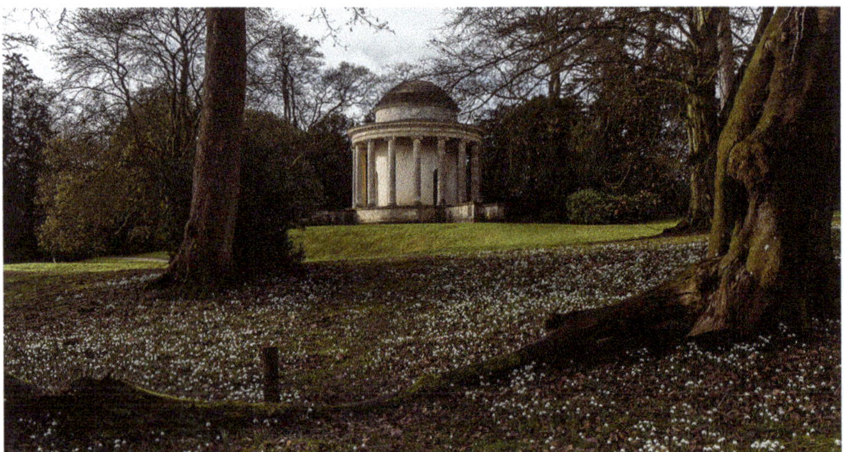

Figure 13. Photo of part of Capability Brown's eighteenth-century, "mimetic" gardens at Stowe, England. The garden uses plants, topography, texts, and structures to represent a range of political, spiritual, and artistic concepts. Reproduction of "Stowe Landscape Gardens" by Hugh Mothersole is licensed under CC BY 2.0.

emotion or imaginative vision suffused with emotion. What an individual artist felt or understood about something was deemed valuable and worth expressing in a work for the personal, emotional, or intellectual benefit of the person reading, viewing, or listening to the work.[47] Monet's famous and much-visited garden at Giverny in France is a good example of an "expressive" garden (see Figure 14).

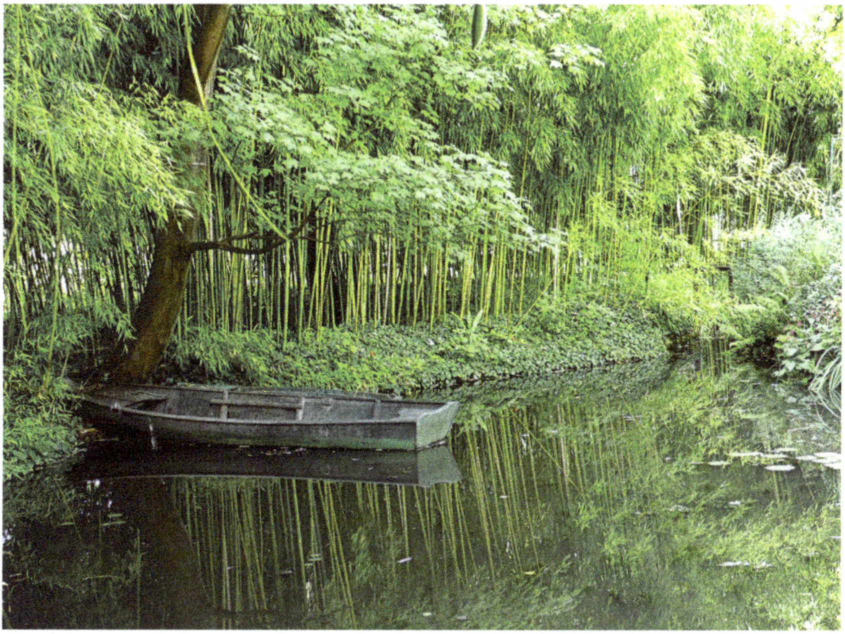

Figure 14. Photo of part of Monet's "expressive" garden, established from the late nineteenth century onwards, at Giverny, France. The garden uses plants, water, light, and structures to express, and thereby communicate, Monet's way of seeing and understanding the world. Reproduction of "Giverny – Monet's garden, lily pond, boats" by Damian Wentwhistle is licensed under CC BY 2.0.

47 See: Graham, "Expressivism: Croce and Collingwood," 156–63; John Spackman, "Expression Theory of Art," ed. Michael Kelly, 2nd edn, *Encyclopedia of Aesthetics* (Oxford University Press, 2014), <http://www.oxfordreference.com/view/10.1093/acref/9780195113075.001.0001/acref-9780195113075-e-0204>. § 1. For good accounts of the expression theory in its nineteenth- and twentieth-century guises see: ibid., Introduction, § 1, § 3.

Modernism eventually succeeded the expression theory as the dominant conception of art for its time, although both conceptions co-existed well into the twentieth century.

Autonomous modernism claimed that art was self-referential and that its value was independent of, and as important as, other values, such as truth or beauty.[48] According to autonomous modernist theories, the raison d'être of art was the creation of works which exhibited aesthetically satisfying forms ("significant form"), disinterested contemplation of which gave rise to an intrinsically valuable aesthetic experience for the perceiver. The image below of part of Thomas Church's pool garden at Sonoma, California, exemplifies well some of modernism's typical preferences, and its ethos (see Figure 15).

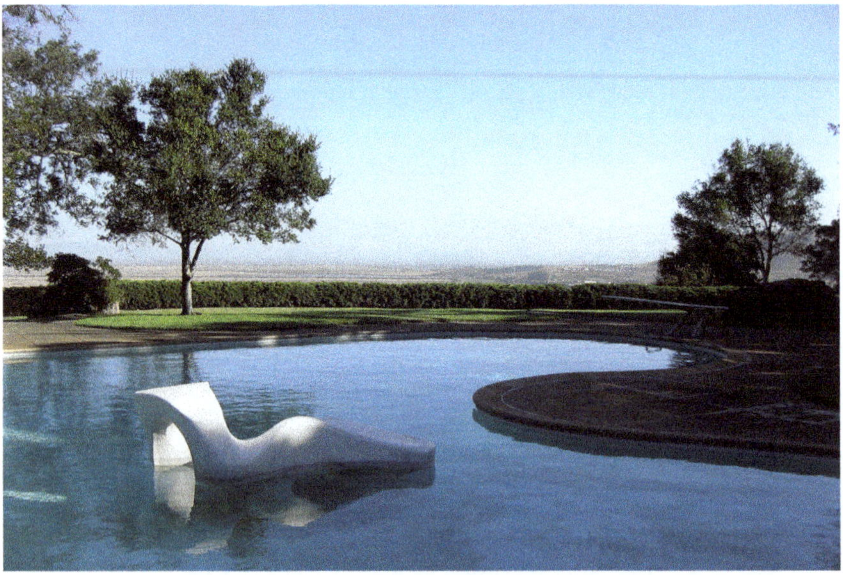

Figure 15. Photo of part of the modernist Donnell Ranch Garden at Sonoma, California, designed by Thomas Church in 1948. The pool area exhibits the clear, precise, unadorned lines, shapes, and volumes preferred, for their intrinsic qualities, by modernist critics. Reproduction of "Donnell garden" by Morisius Cosmonaut is licensed under CC BY 2.0.

48 Modernism also had an important branch that might be termed engaged modernism. I have not found evidence to suggest that engaged modernism had more than slight effects on garden art.

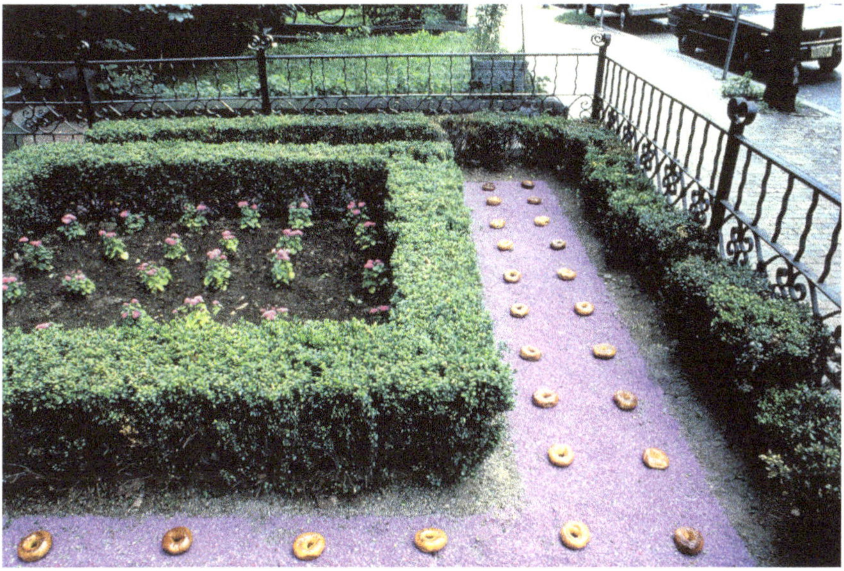

Figure 16. Photo of Martha Schwartz's 1979 *Bagel Garden*, in Boston, MA, USA. The garden's arthood derives not from the artist's selection and arrangement of her materials but, rather, from the ways in which the garden relates to the history of art and to the contemporary artworld. Photo reproduced with permission from Martha Schwartz Partners.

The three conceptions discussed so far all focus on *manifest* properties, or relationships that depend on manifest properties of works of art. The last conception to be considered, the relational theory of art, is concerned only with a work's extrinsic and relational properties. The relational theorist acknowledges something to be art not on the basis of any visual or other manifest properties an object or event may possess, but on the basis of the relationships that object or event enjoys with respect to the history and theory of art, or its relationship to what Dickie christened the "artworld."[49] Martha Schwartz's *Bagel Garden* enjoys a playful interaction with the history of garden art and is presented for appreciation by a member of the contemporary (garden) artworld. Therefore, according to the relational theories, it is "art" (see Figure 16).

49 George Dickie, *The Art Circle: A Theory of Art* (New York: Haven Publications, 1984), 80–2.

CHAPTER 3

Changing Places: Gardens and Philosophical Aesthetics (1700–2017)

In *What Gardens Mean*, Ross takes eighteenth-century English gardens as her paradigmatic case of gardens-as-art or, in her words, "high art."[1] According to Ross, those gardens constituted a fully formed contemporary art form and they and their makers were celebrated, examined, and argued over by contemporary philosophers and writers. In other words, their stake in philosophical aesthetics was high.

Ross's view of the important artistic role gardens played in eighteenth-century life has been supported by many including, most recently, Steven Heyde, author of a 2014 paper that provides important insights into the origins of landscape architecture's conceptions of the aesthetic. In the paper, he writes that the "status of a garden as a worthy form of art was at this point in its history [the eighteenth century] not only proclaimed by a few professionals, but supported by a wider cultural elite of owners, philosophers, poets, architects, and artists." He adds further that "for many of them," the landscape garden was "a unique form of art, if not a superior form of art."[2]

By the end of the eighteenth century, Kant had cast his considerable shadow over aesthetics. Kant classified gardens ("landscape gardening") as a subset of the art of painting, which "consists in no more than decking out the ground with the same manifold variety ... as that which nature presents to our view, only arranged differently, and in obedience to certain forms." And, further, "the beautiful arrangement of corporeal things [i.e.

1 Ross, *What Gardens Mean*, xiii.

2 Heyde, "The Historical Roots of 'Aesthetics' in Landscape Architecture: An Introduction," 139.

landscape gardening] … is also a thing for the eye only, just like painting."[3] On the surface these remarks appear helpful to the status of gardens as art, but a closer reading belies this judgement.

In comparing the art practices of gardens and painting Kant is not saying anything new. Pope had said something similar as early as 1734 when he wrote, "all gardening is landscape-painting; just like a landscape hung up," and a comparable remark is attributed to Pope's celebrated garden designer, William Kent.[4] But, in Kant's case, because of his almost continuous influence over nineteenth- and twentieth-century aesthetics, and because the description appears in that part of his *Critique* where he is explaining with typical thoroughness the different ways in which the various fine arts function, his equation of gardens with landscape paintings can be interpreted as a rejection of the garden as anything other than a static, visual entity. This conception of gardens – a consideration of the inadequacy of which is a thread running through this study – has dominated philosophical aesthetics until the last three or so decades.

There is another remark of Kant's that initially appears helpful to the understanding of gardens as works of art. He writes, "nature proved beautiful when it wore the appearance of art; and art can only be termed beautiful, where we are conscious of its being art, while yet it has the appearance of nature."[5] However, as the nineteenth century advanced, it was the very naturalness of gardens that turned out to contribute significantly to their reduced status as works of art.[6]

Kant's remark on the necessary co-dependence between art and nature differs in intent from those made by earlier writers. Heyde notes that when the garden designer Stephen Switzer, an early practitioner in the English landscape school, wrote admiringly of a garden as a place where "nature is

3 Kant, *Critique of Judgement*, § 45.

4 Quoted in: Katherine Myers, "Visual Fields: Theories of Perception and the Landscape Garden," in *Experiencing the Garden in the Eighteenth Century*, ed. Martin Calder (Oxford; New York: Peter Lang, 2006), 252.

5 Kant, *Critique of Judgement*, § 45.

6 For a discussion of the problems that arose for art gardens in the context of their "new" naturalness, see 68–70.

truly imitated," or, when Shaftesbury encouraged the use of natural features for gardens, such as "rude Rocks" and "broken Falls of Waters," in order to best represent "Nature," it is not the imitation of Kant's "nature" that they had in mind.[7] It was the nature of Nature, in the widest sense, that they were concerned to represent, and the subsequent realignment of mimetic intent incorporated in Kant's writings is a further significant contributing factor to gardens' changed status as works of art in the nineteenth century.

Furthermore, Kant's characterization of the garden as partaking of both art and nature required him to distinguish its appreciation from the appreciation of works of art in other artforms. It required that he "factorizes" his appreciation, that he appreciates gardens *separately* as nature and as art, thereby missing out on an essential characteristic of gardens: the deliberate blurring of the boundaries between and, alternately and simultaneously, the parrying and thrusting between art and nature that gardens continuously enact.[8]

After Kant, the next canonical philosopher to comment directly on gardens was Hegel. He anointed art, not nature, the paradigmatic object of aesthetic interest and it would therefore be reasonable to expect that art gardens would thereby enjoy high aesthetic status.[9] However, Cooper cites him as describing gardening as an "imperfect art," and gardens as being "welcome if they provide 'cheerful surroundings' … [but they are] 'worth nothing in themselves.'"[10] Hegel conceptualized art as the expression of Spirit, and gardens' unavoidable "natural" physicality proved their undoing. According to Salwa, Hegel found gardens "too close to nature and not permeated by the Spirit enough to be regarded as art."[11] After Hegel, I am not aware of any philosophers of aesthetics expressing sustained, significant

7 Heyde, "The Historical Roots of 'Aesthetics' in Landscape Architecture: An Introduction," 125–6.

8 For a comprehensive statement of this view, see: David E. Cooper, "In Praise of Gardens," *British Journal of Aesthetics* 45, no. 2 (2003), 103–7.

9 See: Carlson, "Environmental Aesthetics," § 1.2.

10 Quoted in: Cooper, *A Philosophy of Gardens*, 9.

11 Salwa, "The Garden as a Performance," 376.

views regarding art gardens until the last few decades, during which the renewed interest in them surfaced.[12]

A Changed Status

Around the beginning of the romantic Era, an important change occurred in gardens' art status. This change represents a pivotal "moment" in gardens' transition from art to non-art in the eyes of philosophical aesthetics. The factors influencing the change resist easy categorization and they overlap (frustratingly) in terms of their individual causes and effects. I discuss them below grouped loosely into three categories: changes in philosophical aesthetics; changes in society; and changes in garden fashions and experiences.

Changes in Philosophical Aesthetics

Around the turn of the nineteenth century, three changes occurred in philosophical aesthetics that had important consequences for art gardens. First, there was the change in paradigmatic art objects, as exemplified in the philosophies of Kant and Hegel. Second, there was a change in the conceptual emphasis for art from a mimetic concern with Nature to a concern, when it was interested in nature at all, with nature *tout court*. In gardens, these two changes interacted with the result that gardens became doubly

12 Cooper names Schopenhauer, Wittgenstein, and Heidegger as other philosophers who had something to say about gardens. (See: Cooper, *A Philosophy of Gardens*, 7.) Schopenhauer ranked landscape gardening immediately above his and Hegel's "lowest" art, architecture, and believed that its aim was to promote scenic beauty, which beauty owed more to nature than the artist. (See: Sandra Shapshay, "Schopenhauer's Aesthetics," ed. E. N. Zalta, *The Stanford Encyclopedia of Philosophy* (Summer 2012), <http://plato.stanford.edu/archives/sum2012/entries/schopenhauer-aesthetics/>. Heidegger wrote approvingly of the activity of gardening. (See: Cooper, *A Philosophy of Gardens*, 158–61.)

less art-like. When eighteenth-century gardens left behind their emblematic and painterly associations and became, in the hands of Capability Brown and others, naturalistic, they became in effect abstract art.[13] Whatever else it might have been starting to think at the time, contemporary aesthetics still assumed that art was mimetic – that it imitated the nature of the world. When gardens started to represent "nothing" except nature, their arthood appeared to be diminished and contemporary aesthetics was no longer interested in them, nor did it find it necessary to account for them. Furthermore, while Hegel was promoting art over nature as the paradigmatic object of aesthetic interest he was required at the same time to demote gardens to the status of an imperfect art because gardens are unavoidably natural and irredeemably material; they are not of the Spirit, and therefore, according to Hegel's theory, they are necessarily at best an imperfect art.[14]

As the romantic Era advanced, the third change in philosophical aesthetics to influence the arthood of gardens emerged. It was the change in the dominant *conception* of art. The new conception of art understood art to be an expressive rather than mimetic endeavour, and that conception profoundly influenced all art that was produced under its influence. Gardens were unlikely candidates for art status under the expression theory. Although they had recently largely ceased being mimetic, mimesis had been a role to which they were well suited. They were much less well suited to the expression theory. According to one standard account of it, the theory is centrally concerned with an artist's feelings about an object or event, her infusion of her work with those feelings and, in turn, the reciprocal experience of the *artist's* feeling by the viewer, listener, or reader.[15] All these stages are difficult to imagine in the case of gardens. A garden's continuously unfinished and evolving nature makes it an unlikely conveyor of a fixed set of emotions. A garden is most often the ongoing product of a range of

13 Although Hunt does not go so far as to call Brown's gardens abstract, for some support of this view see: John Dixon Hunt, *The Figure in the Landscape* (Baltimore, MD and London: Johns Hopkins University Press, 1976), 247.

14 Cooper, *A Philosophy of Gardens*, 9.

15 Graham, "Expressivism: Croce and Collingwood," 156–63.

creative and practical minds and hands. The idea that it can transmit the emotional content of a single artist's single experience is unlikely. The temporal and spatial opportunities for experiencing a garden are theoretically infinite. Therefore, unless the artist has foreseen all those limitless possibilities, how can any particular temporal-spatial experience of a garden be known to equate to the artist's preferred temporal-spatial experience and its own ensuing emotional content? For all these reasons, gardens were not well positioned to be categorized as art according to the expression theory, and gardens' art status was thus diminished.

Changes in Society

When the power of absolute monarchy began to lessen in Britain in the aftermath of the Glorious Revolution of 1688 it was accompanied by a gradual, slow increase in the rights and importance of the "ordinary" individual in that society. What was slow and gradual in Britain was more precipitate in France when the French Revolution occurred almost exactly 100 years later. These seminal events set in train movements and further political events that had a profound effect not only on society but also on the arts, including gardens. There are four matters particularly relevant to gardens that have their roots in these societal changes: the rise of the individual artist as a contemporary hero, the rise of the middle class and its enthusiasm for gardens, the popularizing of museums, and changes in the design "profession."

The improving status of the "ordinary" individual, coupled with the decline of royal and aristocratic patronage, encouraged the growth of the perception of the artist as an individual, heroic genius-seer, or, in the memorable phrase of the art critic Lucy Lippard, a "hero and visionary nut."[16] The personal visions such artists produced and the emotions and insights they expressed were deemed valuable by the aristocracy and the emergent

16 Lucy R. Lippard, "Gardens: Some Metaphors for a Public Art," *Art in America* 69, no. 11 (1981), 138.

middle class.[17] However, gardens, in spite of the inspiration and heroics that may be involved, are almost never the unaided work of a single individual.[18] Gardens are usually the evolving product of a team, and they are therefore inimical to the contemporary notion of the individual artist as a contemporary hero whose singular view and personal, emotional insights are to be valued.

Members of the emergent middle class were enthusiastic consumers of art and, while works in some artforms translated relatively easily to the more restricted confines of middle-class life (see, for example, Schubert's songs and piano duets, Wordsworth's poems, and Dickens's novels), gardens did not. The restricted scale of the new middle-class gardens, compared to their princely and aristocratic forbears, and the new horticultural and artistic ambitions of their middle-class owners (discussed below), both contributed to the loss of gardens' status as works of art.[19]

The eighteenth and nineteenth centuries' emphasis on the development of public museums stemmed in part from the notion that middle-class (but not lower-class) people would profit from being able to see great art and other cultural artefacts which previously had not been available to them. In this way, a classification system was subtly erected: what could be displayed in museums was art and what could not was not. By their very

17 The philosopher and historian Isaiah Berlin writes of the romantic artists' "passionate belief in spiritual freedom, individual creativity." Their goals represent "the self-expression of the artist's own unique, inner vision, to set aside which in response to the demands of some 'external' voice — church, state, public opinion, family friends, arbiters of taste — is an act of betrayal of what alone justifies their existence for those who are in any sense creative." Isaiah Berlin, *The Crooked Timber of Humanity: Chapters in the History of Ideas* (London: John Murray, 1990), 57–8. An important philosophical underpinning of the related concept of genius as a necessary component of fine art is to be found in: Kant, *Critique of Judgement*, § 46.

18 See my description of the evolution of Tupare at 32–3.

19 See: F. R. Cowell, *The Garden as a Fine Art: From Antiquity to Modern Times* (London: Weidenfield and Nicholson, 1978), 187; Laurence Fleming and Alan Gore, *The English Garden* (London: Spring Books, 1979), 171–3.

nature, gardens were unable to be displayed and their arthood was in this way diminished.[20]

During the same period, a gradual transition was taking place from amateur art-garden making to the practice of what might be called craft-garden making, and the emergence of the profession of landscape architecture.[21] Although craft-garden making and landscape architecture became dominant in the latter half of the nineteenth century, the seeds of the non-art status of their products were sown at the beginning of the romantic Era. During the eighteenth century and into the early nineteenth century, art gardens had by definition been produced by artists, whether they were professional, like Brown and Repton, or gifted amateurs, like Hoare. But the cloak of the romantic, heroic genius-artist did not sit easily on the shoulders of the later nineteenth-century garden makers, who, as I have explained above, were no longer making "art" anyway.[22] So, as gardens continued to be made throughout the nineteenth century, the making of them was transferred to the craft gardeners, such as Gertrude Jekyll and William Robinson, and, to the extent to which they were interested in them, to the early pioneers of professional landscape architecture, such as Andrew Jackson Downing and Frederick Olmsted.[23]

20 Cooper, *A Philosophy of Gardens*, 8–9.

21 According to Heyde: the first use of the term landscape architect (*architecte paysagiste*) occurred in France at the start of the nineteenth century; during the nineteenth century, what gradually turned into the profession and discipline of landscape architecture took over from the earlier practice and practioners of garden design; that profession was more aligned to the science of horticulture than the art of design; and, garden design became increasingly a matter of applying "rules" of design and style. See: Heyde, "The Historical Roots of 'Aesthetics' in Landscape Architecture: An Introduction."

22 I do not mean to imply the "craft" gardeners such as Gertrude Jekyll and William Robinson were not artists. I mean, rather, that they weren't *soi-disant* romantic artists and therefore, judged by contemporary criteria, they were non-artists. (It may prove interesting to study the rise of female garden artists, from Gertrude Jeykll right through to, say, Penelope Hobhouse and Beth Chatto in recent times, from a gendered perspective. Has the gender of these practioners perhaps worked against their productions being considered "art"?)

23 Turner, *Garden History: Philosophy and Design 2000 B.C.–2000 A.D.*, 244.

Changes in Garden Experiences and Fashions

From the eighteenth century onwards, a growing emphasis on movement through and around gardens became apparent. The experience of gardens in this way was not new. It had been documented from ancient times and has been widely acknowledged.[24] However, there was a renewed emphasis on it on the part of eighteenth-century garden writers and visitors.[25] It was an essential element of an adequate experience of contemporary "circuit" gardens such as Stourhead, where the visitor not only moved from one set-piece tableau to another but was presumably not unaware of the continuously changing visual materials she encountered on the journeys between the set-piece viewpoints.[26] This "new" experience involved experiencing a garden from a continuously changing, self-motivating, unstable viewpoint, or series of viewpoints. However, although there was a renewed awareness of this phenomenon on the part of garden writers, theorists, and visitors, philosophical aestheticians remained unchallenged in their assumption that an appropriate experience of a garden *qua* work of art involved static, stable views of non-dynamic, framed, garden vistas.

Fashions in garden design altered quite markedly between the Enlightenment and the romantic Era. It is hard to untangle the causal

24 For an entertaining description by the Australian philosopher Damon Young of Ancient Greek philosophers walking about in their gardens see: Damon Young, "Philosophy Alfresco," in *Voltaire's Vine and Other Philosophies: How Gardens Inspired Great Writers* (London: Ebury Digital, 2014).

25 See: Linda Parshall, "Motion and Emotion in C. C. L. Hirschfeld's Theory of Garden Art," in *Landscape Design and the Experience of Motion*, ed. M. Conan (Washington, DC: Dumbarton Oaks Research Library and Collection, 2003). See also: John Dixon Hunt, "The Time of Walking," *Studies in the History of Gardens & Designed Landscapes: An International Quarterly* 36, no. 4 (2016). Hunt considers the practice and theory of walking in both historical and contemporary landscape settings.

26 For an analysis of some differing accounts of movement around Stourhead, and of the relative importance and relevance of those accounts to an adequate experience of the garden, see: "Stourhead Revisited & the Pursuit of Meaning in Gardens," *Studies in the History of Gardens & Designed Landscapes* 26, no. 4 (2006), 330–2; Oliver Cox, "A Mistaken Iconography? Eighteenth-Century Visitor Accounts of Stourhead," *Garden History* 40, no. 1 (2012).

relationships between the changes in fashions and the extra-garden factors that (may have) determined them. I have settled on three sets of factors which seem to have had a significant effect of the world of gardens during the nineteenth century: the development of gardens to suit the needs of the new middle-class owners, the rise in the importance of botanical aspects of gardens, and the effects of the new gardens on the experience of gardens.

Members of the emerging middle class became enthusiastic owners of gardens and, in some cases, enthusiastic gardeners. To meet their needs, a range of what might be called *prêt-à-horter* designs was published by enterprising designers in books and magazines.[27] These designs could be installed wholly or in part in average sized suburban lots by gardeners or owner-gardeners. The layouts and styles could be applied to a site willy-nilly, and they generally lacked any "specific cultural resonances" for the client or the public.[28] The gardens that typically resulted from such designs lacked the arthood of their eighteenth-century predecessors. They also lacked the pretensions to expressivity of works in other nineteenth-century art forms. It is, in fact, fair to say that they lacked pretensions to arthood generally (see Figure 17).[29]

There is one aspect of these gardens which on its own was enough to disqualify them from being considered as art: their utility. Utility was anathema to contemporary aesthetics. Whatever changes had occurred in aesthetics over the previous century, disinterestedness was still an essential component of the aesthetic experience. Gardens which included vegetable plots, cutting gardens, spaces for games, etc., such as Loudon and others offered their clients, were unequivocally non-art.

During the nineteenth century, there was a growing movement away from gardens being valued for their emblematic or expressive contents, that is, in effect, for their potential arthood, towards their being valued

27 See, for example, the plans and sketches in: J. C. Loudon, *The Suburban Gardener, and Villa Companion* [electronic resource] (London: The author, 1838), <https://archive.org/details/suburbangardeneooloudgoog>.

28 See: Heyde, "The Historical Roots of 'Aesthetics' in Landscape Architecture: An Introduction," 135–6.

29 Cowell, *The Garden as a Fine Art: From Antiquity to Modern Times*, 201–4.

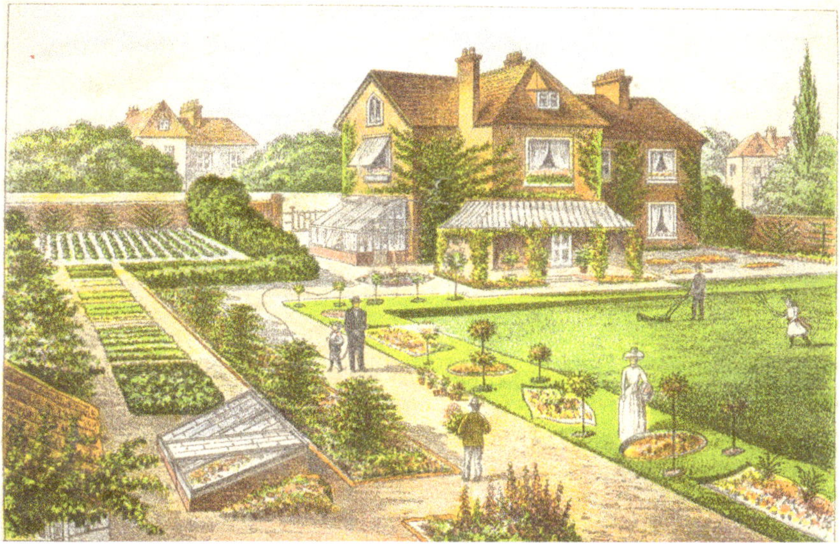

Figure 17. Contemporary advertising image for a typical, off-the-shelf, nineteenth-century villa garden plan, offering potential owners opportunities for satisfying their social, utilitarian, and aesthetic ambitions. The utilitarian features of such gardens were themselves enough to disqualify the gardens from being "art." Image reproduced with permission from Amoret Tanner.

for their botanical content.[30] Although plant hybridizing was not new, it was pursued vigorously during this century and gardeners sought eagerly to install the most novel, biggest flowered, most "improved," most highly scented cultivars in their gardens. Among the most highly prized cultivars

30 Ross proposes two different reasons, both taken from Hunt, why the poetic, allusive garden began to be replaced by less artful gardens, whose aesthetic properties were dependent to a much greater extent than previously on purely visual qualities: first, garden visitors were less well (classically) educated, and therefore gardens' allusive poetic and emblematic content was not understood; second, the influence of "Lockean epistemology[,] with its seeming emphasis on subjectivity and personal response as the groundwork of our knowledge," worked against the appropriate appreciation of the older style, poetic allusive garden. See: Stephanie Ross, "Gardens' Powers," *Journal of Aesthetic Education* 33, no. 3 (1999), 11–12.

were those developed from plant material of exotic origin. Such material was "discovered" on botanical expeditions, which were undertaken by professional botanists and naturalists, and wealthy amateurs alike. The new cultivars produced in this way, as well as the original and other exotic species imported, were able to flourish in their new settings thanks to the increased availability of conservatories, glass houses, and heating systems. The increased availability and efficiency of these structures and systems can be traced to the Industrial Revolution's contemporary stimulation of the disciplines of structural and systems (heating) engineering, and to the growing availability of the necessary materials. These advances in botany and engineering meant that the palette of attractive plants available for use in gardens was much increased (see Figure 18).[31]

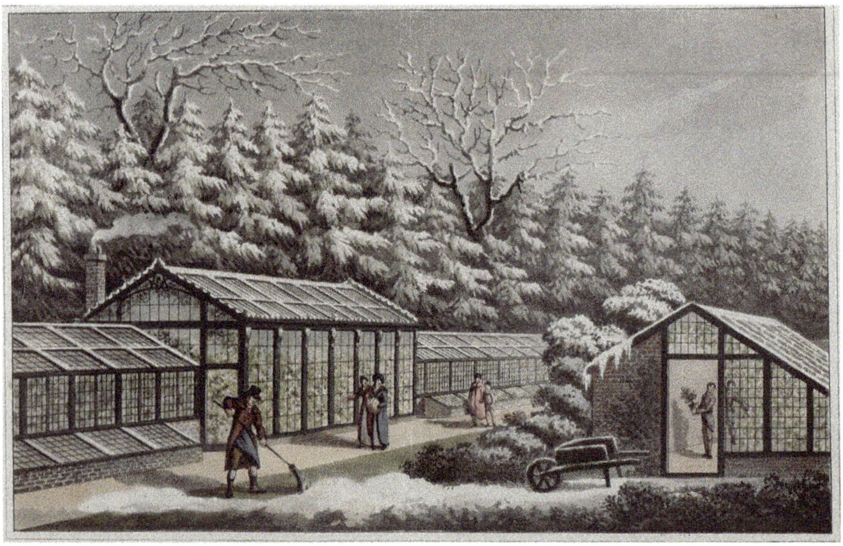

Figure 18. Illustration of Humphrey Repton's Forcing Garden at Woburn Abbey. The 1816 aquatint shows colourful, possibly exotic, plants flourishing in a heated environment under the shelter of a glass and steel structure. The increasing availability of such "over-wintered" plants contributed much to the increased emphasis on botanic novelty in nineteenth-century gardens. Illustration reproduced from Repton's "Fragments on the theory and practice of landscape gardening." ©RHS Lindley Library.

31 Cowell, *The Garden as a Fine Art: From Antiquity to Modern Times*, 193–202.

This increase in plant stock might have been expected to enhance the opportunities for gardens to be considered as works of art. In fact, the reverse was the case. The increased emphasis on botanical variety and the fondness for the novel products of plant breeding programmes meant that individual plants became an important focus of attention in the garden, and this came at the expense of a focus on the whole, or at least a significant part of, the garden as an art object. Furthermore, this focus on the close-up experience of individual plants came at the expense of the contemplation of distant views and objects that had typified the eighteenth-century experience of gardens as art. The heightened focus on individual plants and their aesthetic qualities also came at the cost of the garden's potential for mimesis and expressivity, and in this way too the nineteenth-century garden's arthood was diminished.[32]

One consequence of all the changes that were influenced by botanical innovations was the emergence of a new garden style called the gardenesque (see Figure 19).[33] The gardenesque replaced the picturesque as the dominant garden style. The name change marked a change in the appearance of gardens. Picturesque gardens had, however loosely and with whatever nuances the term might have been applied during the previous 100 years, shared a conception of the garden as in some way being associated with the pictorial arts. On this basis, a garden was artistically successful to the degree that it successfully reproduced aesthetic qualities and formal arrangements found in contemporary pictorial arts, or in nature as seen through art's "eyes." However, the new emphases on botany and horticulture that characterized the gardenesque style meant that a successful garden was to be judged on its botanical and horticultural merit rather than on the old picturesque values. The very adoption of the term "gardenesque" also

32 Landscape architect and garden writer Tom Turner says of this period: "Garden designers … neglected the quest for art to imitate 'the nature of the world.' … They saw only the superficial 'world of nature.'" See: Turner, *Garden History: Philosophy and Design 2000 B.C.–2000 A.D.*, 226.

33 The Oxford English Dictionary's earliest citation for "gardenesque" is from John Claudius Loudon's *Arboretum et Fruticetum Britannicum* of 1838. ("Gardenesque, Adj.," *OED Online* (Oxford University Press, June 2015), <http://www.oed.com/view/Entry/76731>.)

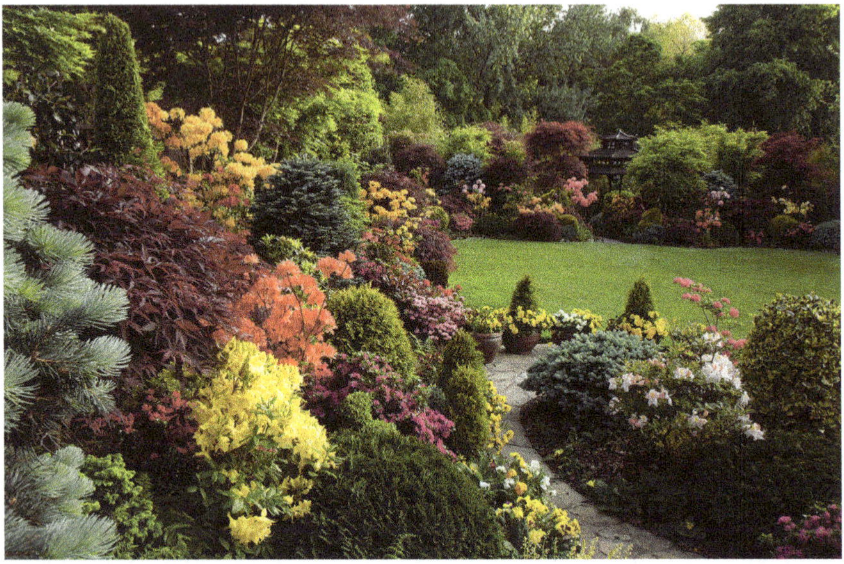

Figure 19. Photo of an extreme, twentieth-century example of a garden in the nineteenth-century gardenesque manner. The overwhelming focus on individual colours, shapes, and especially flowers in this garden undermines any claims it might have to being considered a traditionally conceived coherent artistic entity. Photo reproduced with permission from Four Seasons Garden UK.

necessarily involved the simultaneous rejection of "picturesque," a term that had enjoyed wide currency and that had up to that time been applicable to all the visual arts and to nature. In this way, gardens moved themselves away from the worlds of art and nature appreciation, and the art status they had shared with other arts was consequently diminished.

The gardenesque style resulted not only in gardens that looked different but in gardens that were differently experienced, and these altered modes of experience were inimical to gardens' being considered works of art. At this time, gardeners and garden designers were able to furnish their gardens from a greatly enlarged palette of plants. The resulting *embarras de richesses* was one factor that lead to the increased popularity of planting for seasonal effects and for bedding out, that is, the planting of garden beds in accordance with schemes based on rotating changes of plants selected for their ephemeral aesthetic attractiveness. In support of this style of gardening, Loudon wrote that "one of the greatest of all the sources of enjoyment resulting from the possession of a garden, is the variety which it produces."[34] While the ephemerality and changeability of these garden scenes was well suited to the horticultural displays that the gardenesque style favoured, their instability entailed that visitors could have no fully determined, unchanging experience of gardens containing them. Such gardens necessarily lacked a stable, finished form and were therefore inadmissible as art according to contemporary aesthetics, which continued to require that any work of visual art needed to have an unchanging, fully determined form.

Gardens in the gardenesque style also encouraged an interest in the botanical details of individual plants and in that way offered increased opportunities for the experience of non-visual and, in particular, olfactory sensations. These developments potentially increased the aesthetic enjoyment of gardenesque gardens, but at the same time they diminished the arthood of those gardens because they offered pleasures which ran contrary to the foundational art-aesthetic notion of disinterestedness. It is not possible to appreciate, say, a dahlia with disinterest while admiring it as a product of the hybridizer's skill, or even, to be pedantic about it, while admiring it as a dahlia.[35] Such *interested* admiration lay beyond the boundaries of what was acceptable for the experience of art, at least in the eighteenth and nineteenth centuries. Nor is it possible to enjoy the scent of, say, a rose with disinterest. Disinterest requires a certain distance to be maintained between viewer and art object. Manifestly this is not possible in the case of scent, which must physically enter the body before it is perceived, at which time, disinterest, or psychical distance, is no longer possible. Scent is also an unstable, formless sensation. In Hegel's words, "a smell does not remain stable so that a person can contemplate it" – an assessment with which Kant would have agreed.[36] Therefore, according to contemporary aesthetic theory, enjoying the scent of a rose could play no

34 Loudon, *The Suburban Gardener, and Villa Companion*. 49.

35 Kant, *Critique of Judgement*, § 4.

36 However, see: Tafalla, "Anosmic Aesthetics," 519, 16–20. Tafalla makes a case for smell being an acceptable sensory input in terms of Kantian aesthetics.

role in the appreciation of a work of garden art. Again, any arthood pretensions gardenesque gardens may have entertained were on these grounds diminished.[37]

A Revived Status

In his Foreword to a book written in 2010 by a range of authors on the general topic of gardens and philosophy, Cooper comments on the lack of interest that twentieth-century philosophy, and some other disciplines, have shown towards the garden. He writes:

> The failure, for the most part, of twentieth-century philosophers, cultural historians, and social scientists seriously to attend to the garden was a caesura, a lapse. In earlier centuries, in the traditions of both East and West, the garden occupied an honorable and important place in "the realm of intelligent public discourse" – a discourse engaged in, of course, by philosophers, who had yet to fall victim to the professionalization and specialization that philosophy was to undergo during the last century.[38]

In the Introduction to his own monograph on the subject, Cooper warns that, "in neglecting the garden, philosophy is … ignoring not merely a current fashion, but activities and experiences of abiding human significance."[39]

On a more positive note, the two volumes which Cooper's comments introduce are part of a growing collection of books and papers that have appeared in the last two decades and that are redressing the inaction on the part of philosophers of which he complains. These publications have

37 The problems associated with a disinterested appreciation of nature, and gardens has been well canvassed. See: Arnold Berleant, "The Aesthetics of Art and Nature," in *The Aesthetics of Natural Environments*, eds A. Carlson and A. Berleant (Peterborough, Ontario: Broadview Press, 2004), 76–88; Miller, *The Garden as an Art*, 93–116; Ross, *What Gardens Mean*, 13–14.

38 Cooper, "Foreword," x.

39 *A Philosophy of Gardens*, 3.

appeared in two overlapping "waves." Thomas Leddy, Miller, Ross, and Cooper comprised the first wave, and they have been followed by Salwa, Tafalla, and others representative of an exciting new wave of development in the philosophy of gardens that has been building during the last four or five years.

To understand all this recent activity in the light of the "lapse" of which Cooper speaks it is useful to begin by noting some reasons related to gardens' arthood that may have contributed to the philosophical disdain he detects. Among other reasons, Cooper notes that gardens are "typically and strikingly different … from those artworks which have tended to be regarded as paradigmatic"; that, unlike other works of art, they can be put to "practical, utilitarian uses"; that they lack autonomy because of their "high dependence on environmental factors"; that they are inadmissible according to what Dewey called the "museum conception of art"; and that they are out of tune with contemporary art's concerns and agenda.[40] To this list, Hunt adds they are "antithetical to modernist aesthetics … [because] no garden is ever 'complete' or coherent at any one time."[41]

These reasons have influenced the lines along which the philosophy of (art) gardens developed in the two books with which the renewed period of activity was heralded. In both those works – Miller's *The Garden as an Art* and Ross's *What Gardens Mean* – an important concern was to make a case for gardens' retention in or removal from the category of art.[42] In other words, one of each author's concerns was whether contemporary gardens, in spite of all their differences from other artforms, are or can be art. These art-definitional projects were in tune with the contemporary philosophical aesthetic concern over what constituted a work of art in a *relational* sense, with explaining how something could be a work of art despite its possessing none of the attributes traditionally ascribed to works of art. They were also in tune with the associated comparative disregard for the *intrinsic* aesthetic components of a putative work of art.

40 Ibid., 8–9.

41 Hunt, "Gardens: Historical Overview," 273.

42 Miller, *The Garden as an Art*; Ross, *What Gardens Mean*.

A further important work of this period of revival is Leddy's 1988 paper "Gardens in an Expanded Field." Leddy claims that "gardening [i.e. the garden] has a future as a medium for fine art," but that, if this is to happen, "gardening needs to be discussed in relation to such contemporary art practices as landscape architecture, environmental art, and what has come in the U.S. to be called public art."[43] Gardens do indeed share some features and characteristics with the arts Leddy names, however, given the vast amount of historical written and visual materials concerning gardens and the vast number of gardens in existence, perhaps the practice and critique of these new arts might equally profitably learn from the case of gardens.

The most interesting work now occurring in the field of garden aesthetics does not have its focus on definitional matters. Instead, it focuses on the positive values of the unique, intrinsic components and functions of gardens. In other words, the garden is no longer being examined in order to squeeze it with some difficulty into, or to exclude it from, definitions of art. Instead, gardens are being examined in the light of their own particular aesthetic features and qualities and on account of the unique contributions and insights they may offer to philosophical aesthetics in general. Studies have recently appeared concerning the "performance" of gardens, gardens' temporality, and the extent of gardens' sensory repertoire.[44]

Salwa published his paper, "The Garden as a Performance," in 2013. He argues that our experience of a garden is akin to our experience of other performance arts, such as theatre and music. According to him, the contributors to this experience are the landscape architect or designer who creates the work, nature and art (culture), which play out a continuous dialectical "drama," and the visitors, who are the audience witnessing the "drama."

43 Thomas Leddy, "Gardens in an Expanded Field," *British Journal of Aesthetics* 28, no. 4 (1988), 327.

44 See: "Special Issue on Time," *Studies in the History of Gardens & Designed Landscapes: An International Quarterly* 34, no. 1 (2014); "Sound and Scent in the Garden: Garden and Landscape Studies Symposium at Dumbarton Oaks," <http://www.doaks.org/research/garden-landscape/scholarly-activities/past/sound-and-scent-in-the-garden/abstracts>; Salwa, "The Garden as a Performance"; Tafalla, "Smell and Anosmia in the Appreciation of Gardens."

Salwa also offers an alternative version of the garden *qua* performance, in which he retains the landscape architect as the creator of an audience-less work in which nature, art, and garden visitors are all performers.[45]

In 2014, Tafalla published "Smell and Anosmia in the Appreciation of Gardens," preceded, one year earlier, by "Anosmic Aesthetics."[46] In "Anosmic Aesthetics," Tafalla made claims useful to the development of a realistic account of the experience of gardens. She argued that philosophical aesthetics has traditionally favoured sight and sound over taste, touch, smell, and kinaesthetic experience; that, according to traditional aesthetics, it is not possible to smell with disinterest and therefore smells cannot be part of a traditionally defined art experience; and that the aesthetics of the everyday has given a new, and appropriate, aesthetic status to the sense of smell. In her 2014 paper, Tafalla does not go so far as to agree with Kant that the proper experience of a garden *cannot* involve attention to smells but she does agree with him that the proper appreciation of a garden necessarily involves paying attention to its form. She then goes on to argue that smell is indeed a contributory factor in our appreciation of form and it is therefore an appropriate and necessary sense to employ in the proper appreciation of gardens.

Tafalla's claim that olfaction can be a valid component of the traditionally conceived notion of aesthetic appreciation of form is welcome and her argument opens up an opportunity for others to enlist the other non-visual and non-aural senses onto the roll of "valid" receivers of "valid" aesthetic information in works of art.[47]

45 On pp. 166–69, I give reasons why performance is an inappropriate metaphor for the experience of gardens and an unhelpful concept for explaining their ontological status.

46 Tafalla, "Anosmic Aesthetics"; "Smell and Anosmia in the Appreciation of Gardens."

47 Tafalla is not alone in her philosophical interest in gardens' non-visual pleasures. (See, for example: Miller, *The Garden as an Art*, 32, 47–8; Ross, *What Gardens Mean*, 156–63.) And nor is she alone in paying philosophical attention to the aesthetic values of the olfactory sense. (See, for example: Larry Shiner, "Art Scents: Perfume, Design and Olfactory Art," *British Journal Of Aesthetics* 55, no. 3 (2015).) However, Tafalla is unique in proposing a positive philosophical account of how non-visual,

Beyond philosophy, recent publications in other disciplines have also taken a serious interest in traditionally neglected aspects of the garden. In a 2014 issue of *Studies in the History of Gardens & Designed Landscapes*, non-philosopher scholars considered the functioning and roles of time and temporality in the design, experience, and assessment of gardens and designed landscapes, and, in *Sound and Scent in the Garden*, scholars considered the wide range of sensory opportunities that gardens provide from a range of disciplinary perspectives.[48]

These recent publications are all attempting to capitalize on gardens' unique aspects, to treat those aspects as the feathers in gardens' artistic cap, so to speak, rather than as impediments to gardens' being considered fully fledged works of art. I believe that these features of gardens, including their special ontological, temporal, and experiential characteristics, need to be adequately accounted for in any useful twenty-first-century philosophical account of gardens.

---

in her case, olfactory, sensations in the garden function in the context of received (Kantian) philosophical accounts of art.

48 "Special Issue on Time"; D. Fairchild Ruggles, ed., *Sound and Scent in the Garden*, Dumbarton Oaks Colloquium on the History of Landscape Architecture (Harvard University Press, 2017).

PART III

The Lie of the Land: A Critique

CHAPTER 4

Up the Garden Path: Some Recent Writing

Three substantial books examining art gardens, and gardens and gardening, have appeared during the last 25 years. Their philosopher-authors have each made important claims with respect to art gardens. These include Ross's claim that gardens "are dead in the sense that artists today no longer produce major works in [this genre]"; Miller's claim that "[i]f gardens are works of art, as it now seems they are, they are so in spite of the fact that they do not fit our definitions of art";[1] and Cooper's claim that the art status of gardens is philosophically uninteresting because the questions it raises are "too close to similar and familiar ones asked about other artworks to raise [any] novel issues." These claims, and others Ross and Miller make, merit detailed examination.

Gardens and Meaning

In her 1998 study, *What Gardens Mean*, Ross uses eighteenth-century England's so-called English landscape gardens as case studies for many of the philosophical claims she makes.[2] These gardens provide her with rich

1 Miller, *The Garden as an Art*, 178; Ross, *What Gardens Mean*, 193. Cooper's fine book, *A Philosophy of Gardens*, cannot be ignored. However, its focus is not on art gardens but on all gardens, and gardening. Cooper claims that the art status of gardens is philosophically uninteresting because the questions it raizezs are "too close to similar and familiar ones asked about other artworks to raise [any] novel issues," a view which I necessarily reject. (Cooper, *A Philosophy of Gardens*, 12.)

2 Ross, *What Gardens Mean*.

source material for her principal claims, that: gardens were once a "high art"; gardens can function in the manner of poems; gardens can function in the manner of paintings; gardens offer a distinctive range of meanings and experiences; and that the art of garden making is now dead.

Ross's first claim, that eighteenth-century landscape gardens were considered high art by contemporary writers and philosophers, appears irrefutable. She provides ample written evidence to support this claim and uses a quotation from the eighteenth-century historian and man of letters Horace Walpole to reinforce her point and introduce the relevant section of her book: "Poetry, Painting, and Gardening, or the Science of Landscape, will forever by men of Taste be deemed Three Sisters, or *The Three New Graces* who dress and adorn nature."[3] Further, she provides evidence to show that garden makers such as "Capability" Brown, Humphry Repton, William Kent, and Henry Hoare, and gardens such as Stowe, Stourhead, Rousham, and Chatsworth, were celebrated at that time as, indeed, the gardens continue to be today.

Ross's second claim, that gardens can function in the manner of poems, leads her to consider the adequacy of gardens to the tasks Aristotle sets down for poetry in the *Poetics*. "In particular," she wants to know "how can gardens deal with great human subjects and represent significant human actions?" and how gardens may function as meaningful entities.[4] Many important eighteenth-century gardens, such as Stowe and Stourhead, were laid out as pedestrian circuits, progress around which lead visitors through quite complex iconographical systems, and an experience of such gardens was sometimes described by contemporaries as a "reading" of the garden.[5] According to Ross, these iconographical systems comprised views, scenes, benches, inscriptions, sculpture, architectural ensembles, and monuments (but not, I note, trees, shrubs, or flowers). Progress around such gardens was "poetic" because the iconographic components were experienced in a

3 Ibid., 49.
4 Ibid., 50.
5 Ibid., 51.

pre-ordained temporal sequence and because the components themselves were semantically or otherwise referential.[6]

Hunt writes that the garden at Stowe offered "constant examples of visual exhibits, often accompanied by inscriptions or mottoes, the full meaning of which depends upon the exact encounter of word and image that we find in the emblem book."[7] Emblem books "were intended to assist the inspiration of poets, painters, and orators, and their association of word and image soon became common coinage. Educated people, seeing a particular phrase or image, would know its traditional meaning and associations."[8] The meaning and associations were typically drawn from "classical culture, the Bible, and more primitive lore and superstition."[9]

So, for Ross, an eighteenth-century garden's referential meanings are conveyed in one or both of two ways: first, words appearing on tablets, monuments, and so on are referentially meaningful *qua* words, and second, visual images, such as those recognized through familiarity with emblem books, are referentially meaningful *qua* visual images; and gardens containing such references are poetic because their words and their emblematic visual images are to be experienced in a pre-ordained temporal sequence.

Granted a licence acceptable during the pre-Laocoönian period of which she writes, Ross has shown that gardens of that time can be seen to function in the manner of poems even though the mechanism of visual representation which that involves, in the case of the emblems, belongs to painting and sculpture. And this of course is correct. But I am interested in what makes gardens *distinctive* as an art, and Ross's implied conclusion that gardens are a type of *poème-en-plein-air*, in which plants – surely the most distinctive and characteristic component of gardens – are not necessarily relevant, does not help explain that distinctiveness. However, her

6 For an interesting account from a literary perspective of temporality, and taste in the garden at Stowe see: James Noggle, *The Temporality of Taste in Eighteenth-Century British Writing* (Oxford: Oxford University Press, 2012), 67–95.

7 Quoted in: Ross, *What Gardens Mean*, 51.

8 Ibid.

9 Ibid., 54.

reinforcing of the importance of the sequential element in the experience of gardens is to be welcomed.

Ross's third claim is that gardens can function in the manner of paintings. She describes three ways in which this can come about: "a garden can recreate or copy a landscape painting; a garden can allude to, evoke, recall, or remind us of a painting; and ... a garden can *function* as a landscape painting if it represents some other piece of land, either real or ideal."[10] Certainly, in the eighteenth century, gardens were often created with one or more of these aims in mind, and this should not be surprising because then, as now, garden making, in the professional, philosophical, and popular minds, was commonly associated with notions of pictorialism. That connection was especially strong in the eighteenth century, when the design of gardens was much influenced by the experiences garden owners had while on their "grand tour." In this way, gardens came to be created which reflected ancient sites, complete with exotic trees and ruins and, equally, they were created to imitate paintings, such as those by Claude and Poussin, which in turn reflected real or imagined ancient sites.[11]

According to Ross, these painterly gardens were designed primarily to facilitate visitors' sensory, non-intellectual experiences and to satisfy their sensory, non-intellectual needs. Writing of Charles Hamilton's famed painterly garden at Painshill, she says that it "offered visitors a series of engaging visual scenes with contrasting emotional tones and carefully composed visual surprises, but it did not have a complex meaning that visitors were to puzzle out."[12]

Ross's position regarding gardens that function in the manner of paintings is part of a well-established tradition, both in the scholarly literature and in folk traditions. However, that tradition tells only part, and, for me, not the most important part, of the story about gardens. Gardens are not two-dimensional or even three-dimensional works of art, they are in

10 Ibid., 91.

11 Ibid., 91–3.

12 Ibid., 87.

fact four-dimensional arts and therein lies their distinctiveness and their particular aesthetic interest and values.[13]

Conceptualizing gardens as paintings again downplays the importance of plants in gardens. It is a conception that denies plants a presence as living, changing, unavoidably four-dimensional objects, thereby reducing their role to that of static contributors to an unchanging scene.

Ross's fourth claim is that gardens offer a distinctive range of meanings and experiences.[14] In support of this claim, she refers to the philosophy of art of the twentieth-century American philosopher, Susanne Langer who, in her 1953 study, *Feeling and Form*, developed a general philosophy of art and detailed philosophies of eight artforms and some sub-artforms.[15] Her list of artforms included architecture, but not gardens. She defined art as "the creation of forms symbolic of human feeling."[16] By this she did not intend to claim for art anything like that claimed by nineteenth-century expression theories. Rather, she meant to claim that an artist expresses "what he *knows about* the so-called 'inner life'" [my emphasis], and not his own emotions, in a work.[17]

The means by which this knowledge about the inner life was transmitted was unvarying across the arts. It consisted in the "logically expressive, or significant, form," which was to be apprehended in the "primary illusory field" of each artform. In painting, for example, the form was to be apprehended in the virtual three-dimensional scene created on a two-dimensional canvas and, in architecture, the form was to be apprehended

13 I return to this theme throughout Chapter 5.

14 Not all philosophers agree. For contrary views see: G. R. F. Ferrari, "The Meaninglessness of Gardens," *The Journal of Aesthetics and Art Criticism* 68, no. 1 (2010); Jane Gillette, "Can Gardens Mean?," in *Meaning in Landscape Architecture and Gardens: Four Essays; Four Commentaries*, ed. Marc Treib (New York: Routledge, 2011).

15 S. K. Langer, *Feeling and Form: A Theory of Art* (London: Routledge and Kegan Paul, 1953).

16 Ibid., 40.

17 Ibid., 28.

in the virtual ethnic domain, or image of a culture, formed by the architect's created space.[18]

Langer's significant, or living, form involves the creation of a symbol of our felt life.[19] This symbol does not express an artist's emotions, nor does it stimulate emotions in the perceiver. Rather, the living symbol mirrors aspects of our inner life and, accordingly, we are enriched by coming to know more about that life. In all the artforms Langer talks of, this symbol is presented complete and unchangeable at the time the work is finished. In the case of singular works, like *Mona Lisa*, the canvas contains the unchanging living form which viewers detect and experience. In works for performance, like *Messiah*, the score provides more or less accurate instructions for a successful presentation of the unchanging living form, which is brought to life by the performers and detected and experienced by listeners in any adequate performance of *Messiah*. But Langer offers no way in which a living form may be incompletely or inadequately presented as a work of art in the way in which a garden making artist inevitably first leaves her work. Nor can her living form accommodate the myriad of changes which, intentionally or otherwise, inevitably occur in gardens that may endure for centuries. In other words, Langer's theory cannot account for works that are ontologically incomplete and to which our epistemic access is also incomplete. Nor does she envisage works, such as gardens, which not only change but which may, unlike, say, painting and sculptures, go out of existence of their own accord. Therefore, Ross's use of Langer's theory to support her own claim, that the living form of a garden is to be apprehended in that garden's virtual world, is inappropriate.[20]

Ross also claims that Langer's concept of a work's illusory field is a useful basis for arguing that although a garden is inevitably real, a garden also needs to function as an illusory field if it is to be art. Ross bases her

18 For Langer's discussions of painting and architecture see: ibid., 69–88 and 92–102 respectively.

19 Langer also described significant form as living form and I will use this term from now on to avoid confusion with the early twentieth-century art theorist Clive Bell's significant form, from which Langer's differs in important ways.

20 Miller makes similar use of Langer's theory of art in *The Garden as an Art*.

claim on an extrapolation she makes from Langer's theory of architecture. Langer conceptualizes architectural buildings as if they are stand-alone objects, which, with appropriate detachment, we experience *qua* art. But, this is not generally how we experience architecture and, in this way, our experiences of architecture, and gardens also, are importantly different from our experiences of all the other arts. In the case of novels, film, paintings, symphonies, and so on, it is *relatively* clear what constitutes the work, what its limits are, what an appropriate experience of it consists in and how much of it we need to see or hear or read before we can say we have experienced the work. However, this is not the case in architecture and it is certainly not the case in gardens. The spatial extent of gardens is both uncertain and, typically, large, and their temporal extent is open ended. How do we know if we have seen the (whole) garden? In what order, and therefore via which transitional zones, should we experience the garden? Do we need to see it in four seasons? At what stage(s) of growth should we see the garden? Should we see the garden by moonlight? Should we *smell* the plants? And so, acknowledging the existence of all these possibilities, how are we to know whether we have apprehended the "correct" version of the architect's illusory ethnic domain or the garden maker's garden's virtual world? Langer offers no solution to this dilemma. Her understanding of how architecture is experienced is, on this basis only, faulty, and any extrapolation from it to the case of gardens is problematic.

There remains a third way in which Ross's use of Langer's philosophy is unhelpful to her project. Ross makes important claims in her book regarding gardens' ability to function as both poems and paintings; but these claims are not consistent with Langer's philosophy, according to which each art has a *unique* illusory field through which it conveys the living form that is constitutive of its meaning. For Langer, the idea of gardens as poems or paintings is unacceptable. This does not mean that Ross's treatments of these topics are invalid. They are indeed interesting and informative. However, a question may reasonably be raised as to the appropriateness of Ross's invoking Langer's philosophy of art to support part of her own argument while a key element of Langer's philosophy is at the same time contravened, or at least ignored.

To build on Langer's concept of the garden as a virtual world, Ross goes on to invoke concepts taken from the contemporary philosophers Richard Wollheim, Arthur Danto, and Nicholas Wolterstorff.[21] From Wollheim she takes the idea of "twofoldness," from Wolterstorff the idea of the "world of the work," and from Danto the idea of the "material counterpart" of a work of art.[22] She is then in a position to make her final claim in this regard, that "gardens are simultaneously physical and virtual worlds and are experienced as such by us." And further, "though we may each have different thoughts and feelings as we stroll through a given garden, we enter a shared virtual garden, one composed of the maximally compossible set of experiences elucidated and extrapolated from that physical realm."[23] I agree with her that our experiences of a garden involve our interactions with the "maximally compossible set of experiences" a garden has to offer but I disagree that that necessarily entails the garden world's virtuality. My preference is that the garden's physicality be celebrated and grappled with and I develop this theme in the final chapter of this study.[24]

Ross's fifth claim is that the art of garden making, along with tapestry and stained glass, is now dead.[25] She writes that "all three arts … are dead in the sense that artists today no longer produce major works in these genres," and that "we do not consider gardening a high art, and artists do not make major statements in this medium."[26] This, I argue, is not the case.

The evidence for the arthood of gardens since the turn of last century is extensive and, I claim, hard to refute. A roll call of garden makers that includes architects Le Corbusier, Richard Neutra, Charles Jencks, Thomas Church, and Gunnar Asplund, filmmaker Derek Jarman, and garden designers and landscape architects Geoffrey Jellicoe, Roberto Burle Marx, Brenda Colvin, Russell Page, Christopher Tunnard, Martha Schwartz, Ian Hamilton Finlay, Fernando Caruncho, and Gilles Clément,

21 See 181–4 for my discussion of the theories of these three philosophers in the context of gardens' materials.

22 Ross, *What Gardens Mean*, 178–86.

23 Ibid., 186.

24 For Cooper's rejection of garden as virtual worlds see: Cooper, *A Philosophy of Gardens*, 18.

25 Ross, *What Gardens Mean*, 192.

26 Ibid., 193, xii.

to name but a few, cannot be lightly dismissed as a reason for considering garden making an alive rather than a dead art. Ross herself cites two names from this list when she refers to Martha Schwartz's *Stella Garden* and Ian Hamilton Finlay's *Little Sparta*.[27] This being the case, it seems that, for Ross, a certain, unspecified amount of activity may be necessary before a moribund art can in fact be considered vital.

Perhaps, given the evidence just presented for contemporary garden making's artistic vitality, and given garden making's well documented contemporary "folk" vitality, this claim of Ross's – that garden making is a dead art – should rather be that garden making has been, until most recently, a dead or unfashionable art for *philosophers*, and that that is the case because, for whatever reasons, almost all contemporary philosophers have chosen to ignore it.

It is also hard to accept Ross's concomitant claims that tapestry and stained glass are dead arts. As I write, one of New York's foremost dealer galleries – and part, surely, of any artworld – has advertised for sale tapestries created after designs by Lichtenstein, Le Corbusier, Chagall, Stella, Picasso, and many others.[28] And, although stained glass production is now at a comparatively low level, probably because of its traditional association with places of traditional worship, artists continued to produce work in this artform as the twentieth century progressed. Once again, a roll call that includes artists Marc Chagall, John Piper, Henri Matisse, and architect Frank Lloyd Wright, and extends to the work of the important contemporary artist Olafur Eliasson cannot easily be disregarded.[29]

The point I wish to make here is not that Ross is incorrect in making these claims about "dead" arts. Perhaps, as I have already suggested, it is just a matter of degree, and Ross may require a higher threshold than I do for assessing whether an art form is vital or not. However, by arguing that

27 Ibid., 206–7.

28 See: "Jane Kahan Gallery," <http://janekahan.com/tapestries/> [accessed 26 August 2016]. See also the retrospective exhibition of twentieth-century tapestry mounted at Kunst Haus in Vienna in 2000: "Masters of the 20th Century." <https://www.kunsthauswien.com/en/exhibitions/archive/38-2000/136-tapisserie>.

29 For an example of Eliasson's stained glass work see: Olafur Eliasson, "Your Rainbow Panorama," <http://en.aros.dk/visit-aros/the-collection/your-rainbow-panorama/>.

gardens are a dead art she is able to proceed to an important conclusion of her book: "some arts die when supplanted by more vigorous successors."[30] In the case of gardens, she claims these vigorous successors are environment art and land art. I do not contest Ross's claim that environment and land art are (comparatively) vigorous forms of art today, but these arts may be understood simply as new arts, and not as something into which gardens have evolved by "an artistic version of natural selection."[31] The emergence of these new arts art may be linked not to the morbidity of gardens but to a range of changes in art and society that have engendered a plethora of new arts, such as conceptual art, performance art, installation art, computer art, street art, and so on. If, however, Ross's claim is correct and the eighteenth-century landscape garden *has* evolved into the environment and land art of today, then her claim would be strengthened is she were to offer some account of what happened to gardens between the end of the dominance of the English landscape school and the emergence of environment and land art approximately 200 years later.[32]

Gardens and the Definition of Art

One of Miller's principal concerns in *The Garden as an Art* is the adequacy of what she calls "aesthetic theory" to the concept of art.[33] Miller investigates this by examining George Dickie's definition of art and Munroe Beardsley's theory of art and concludes that Dickie's definition and Beardsley's theory

30 See: Ross, *What Gardens Mean*, 193.

31 Ibid.

32 In a 2006 paper focusing on the appreciation of nature, Ross appears to reject some of the claims she made about gardens in *What Gardens Mean*. However, in the absence of supporting arguments, it is not possible to evaluate these revised claims. (See: "Paradoxes and Puzzles: Appreciating Gardens and Urban Nature," *Contemporary Aesthetics* 4 (2006).)

33 Miller, *The Garden as an Art*, 5.

are both inadequate to the case of gardens. She argues that Dickie's definition is inadequate because it does not allow for the demotion of art gardens from their eighteenth-century pinnacle to what she claims to be their non-art status of today. Her rejection of Beardsley's theory is less straightforward: she claims that gardens, having experienced a twentieth-century "fall from grace," are no longer art, although, at least in the USA, they are "aesthetic," and, therefore, Beardsley's theory should account for them, but it doesn't because they are not thought of as art. Therefore, Beardsley's theory is inadequate because gardens, although aesthetic, are not considered to be art, and because the theory "does not adequately catch our practical de facto definition of art."[34]

Miller's interest in definitions is understandable because, during much of the historical period discussed in the previous chapter, definitions of art were not equipped to deal with gardens or, if they did deal with them, they did so only by ignoring gardens' unique qualities. Miller's claims with respect to Dickie's definition and Beardsley's theory constitute an important part of her book.[35] However, these claims tell the reader more about definitions (and theories) than they do about gardens. Whether or not a garden is "art," according to a given definition, involves satisfying some *de minimis* condition for what is "art" and is not necessarily helpful in clarifying what is characteristic, valuable, and unique about gardens.

At different points in her book, Miller offers differing conclusions in respect of her overarching claims regarding the contemporary art status of gardens. For example, she writes: "Ironically, the garden has to be rejected as an artkind precisely on the theory designed to be the most generous and inclusive – George Dickie's 'institutional' definition of art."[36] And, "I have argued that gardens might well be considered an artkind ... because they do ... fit such current definitions of art as those of Munroe Beardsley and George Dickie."[37] And again, "If gardens are works of art, as it now seems

34 Ibid., 70.

35 Miller uses theory and definition coterminously. I discuss problems stemming from this at 102–3.

36 Miller, *The Garden as an Art*, 70.

37 Ibid., 121.

they are, they are so in spite of the fact that they do not fit our definitions of art."[38] However, I believe a fair assessment of her overall position to be that she rejects the adequacy of both Dickie's and Beardsley's accounts when tested against the case of gardens[39]

Dickie's definition of art exists in more than one version and Miller refers to the earliest version of it in her text. She writes of it that "something becomes an artwork if 'some person or persons acting on behalf of a certain institution (the artworld)' designate it 'a candidate for appreciation.'"[40] Had Miller chosen to engage with the later, 1984 version of Dickie's definition, which was well known at the time of the publication of her book, her conclusions *may* have been different. In its later version, Dickie's definition lays out five interlocking "sub-definitions" of (1) an artist, (2) a work of art, (3) a public, (4) the artworld, and (5) an artworld system. I quote this version now in full because a close reading of it supports the claims I make below regarding it and its inclusiveness to the case of gardens. Dickie's sub-definitions read:

> An artist is a person who participates with understanding in the making of a work of art.
>
> A work of art is an artefact of a kind created to be presented to an artworld public.
>
> A public is a set of persons the members of which are prepared in some degree to understand an object which is presented to them.
>
> The artworld is the totality of all artworld systems.
>
> An artworld is a framework for a presentation of a work of art by an artist to an artworld public.[41]

38 Ibid., 178.

39 Miller's view of the arthood of gardens had changed by the time of her "Gardens: Gardens as Art" of 1998, but that text offers no reasons to explain or support her changed views. She writes that "Munroe's Beardsley's generous principle [for assessing arthood] would clearly include gardens." And, "similarly, George Dickie's 'institutional definition' … would also accommodate gardens." See: "Gardens: Gardens as Art," in *Encyclopedia of Aesthetics*, ed. Michael Kelly (New York, Oxford: Oxford University Press, 1998), 277.

40 *The Garden as an Art*, 70.

41 Dickie, *The Art Circle: A Theory of Art*, 80–2.

Note that Dickie defines the artworld as "the totality of all artworld systems." This is important for my argument and leaves the door open for the claims I make regarding subsets of the artworld, and artworlds that may run in parallel to *the* artworld, should such an institution exist.

Miller makes two related claims regarding Dickie's definition.[42] First, she claims that the definition is inadequate because it cannot account for what she understands to be the downgrading of gardens from art to non-art status. Second, she invokes the "inadequate" definition to claim that gardens are not now art because members of the contemporary artworld do not present them as candidates for appreciation.

Miller's first claim concerns the adequacy of the institutional definition to accommodating the changing status of objects from art to non-art. She shares the widely held belief that, during the eighteenth century, certain European gardens were considered to be indisputable works of art by philosophers, artists, and the limited number of the public who had access to them. She argues that because gardens no longer hold this pre-eminent, or, indeed, any position in the world of art, Dickie's definition is thereby inadequate because it does not allow for the demotion of an object or art-form from art to non-art status. Her claim makes sense and I agree with her as far as she goes. If she is right in saying that gardens are now non-art then she is also right in criticizing the definition for not having an escape clause, as it were, whereby art could stop being art. But, I don't believe that it is the case that gardens are no longer works of art in terms of Dickie's definition and, consequently, if the definition is flawed in this regard, it is not so because of the case of gardens.

Miller's second claim is related to her first one. It says that gardens are not art because members of the artworld ignore them and do not present them as candidates for appreciation. There is an unfortunate link between her two claims: claim one is that the institutional definition is inadequate

42 In the discussion that follows, I will refer to the earlier version of Dickie's definition so as to be consistent with Miller's use of it. I believe that what is made explicit in the later version is already implicit in a thorough understanding of the earlier one and therefore, while my refuting of Miller's claims may be enhanced by an understanding of Dickie's second theory, it does not need that theory to substantiate it in the first place.

because it cannot accommodate gardens' demotion from art status and claim two uses the inadequate institutional definition to argue that gardens are not art because they have, among other things, been demoted. I reject this second claim, on the grounds that it does not reflect accurately the reality of the contemporary artworld.[43]

The institutional definition of art is a procedural definition, that is, an object's or event's status as art depends not on qualities possessed by the object or event but on its art status being conferred on it my members of the artworld. In this way, the institution of art is similar to other institutions, such as marriage. A man and woman achieve the status of being married not because of any particular features they might possess but because that status is conferred on them by members of an appropriately sanctioned institution, such as a church or a court of law.

In the case of marriage, a partnership may be dissolved by a member of an appropriately sanctioned institution and such may be the case for works of art in the limited cases that I describe below. Also, and more significantly, membership of conferring institutions inevitably changes over time, the preferences and requirements of conferring institutions change, institutions may have culturally divergent views of what constitutes marriage, and society may force changes in the conferring institutions. For example, members of some "marriageworlds" confer the status of marriage on bigamists and others do not, and members of others used to but no longer do. Members of some "marriageworlds" confer the status of marriage on same-sex couples and others do not, some do on parent-arranged marriages and some do not. Changes such as these in the membership, procedures, and preferences of the "marriageworld" surely have an equivalence in the membership, procedures, and preferences of any artworld.

In the case of the alleged demotion of gardens from art to non-art status, just as is the case of marital divorce, there are certain grounds on which the revoking of a previously conferred status may be appropriate and legitimate. In the case of divorce, the status of marriage may be revoked, for example, on the grounds that the groom turned out to be the identical

43 See 88–9.

twin of the intended groom, or on the grounds that a partner was mentally unstable at the time she made her commitment and the status of marriage was conferred. In the case of gardens, there may be reasons for which the status of garden may be legitimately revoked according to the institutional definition. Two such reasons seem possible. First, it may be case that members of the art world were originally mistaken in presenting gardens as candidates for appreciation because they did not really understand what gardens were and they presented them mistakenly. Or, second, it may be the (unlikely) case that gardens originally presented as candidates for appreciation were presented in good faith but they later turn out to be forgeries and, therefore, they need to be withdrawn as candidates for appreciation. In both these cases, gardens that were previously art can become non-art in a way acceptable to the institutional definition. However, neither of these reasons applies to eighteenth-century landscape gardens and therefore they, and all "art" gardens thereafter, ought still to retain their art status.

But, Miller's second claim argues that gardens are not art anyway because members of the artworld, as she understands it, do not present them for appreciation. In making this claim, Miller is treating the artworld institution as an unchanging, ahistorical, catholic institution when in fact it is similar in its dynamism to many other institutions in society and frequently revises or changes its membership, procedures, and preferences. Instead of gardens not being art, it may simply be the case that membership of the artworld has changed in such a way that gardens, although they remain art, are no longer of interest to current members of the artworld. Perhaps the new membership has different interests and priorities. Or, perhaps gardens themselves have changed, or not changed, in ways that make them no longer of interest to the current artworld. Or, perhaps (some) members of the artworld may be ignorant of contemporary garden-making for a range of reasons, such as geography or lack of reproductions, or because they harbour, say, a social prejudice against them.

Miller makes no allowance for institutional change and therefore she ends up assessing Dickie's theory as inadequate to the case of contemporary gardens on the grounds that members of the artworld are not presenting gardens as candidates for appreciation. Her assessment is mistaken because, although contemporary gardens do not command the stellar art status of eighteenth-century gardens, it remains the case that throughout the twentieth century, and continuing up to today, there continues to be a thriving "artworld" whose members enthusiastically present for appreciation, discuss, influence, curate, write, research, and lecture about contemporary gardens. This "artworld" has its own procedures, trends, preoccupations, conceptions, heroes, scandals, and so on. I believe the only question to be debated is whether that garden-artworld is part of *the* artworld, or whether it is an artworld on its own account. According to Dickie, such a question is irrelevant because "the artworld is the totality of all artworld systems." However, whether the question is deemed relevant or not, whichever way it is answered, it will still entail that contemporary gardens are art.

Garden writer and sociologist Michel Conan answers that question by asserting strongly that the garden artworld wishes to be considered a quite separate entity from *the* artworld. In his "Introduction: In Defiance of the Institutional Art World," he writes, "garden art deserves scholarly scrutiny in its own right and should be studied in its own terms without any pretence at imitating critical discussions of the contemporary art world since it has remained alien to its critical discourse."[44] Others, such as Miller, Ross, and Cooper, take a more conciliatory approach and, even while they argue about whether gardens are art, they are, de facto, placing gardens within the sphere of interest of *the* artworld.

It is true that the avant-garde artworld, which some mistakenly identify with Dickie's artworld, has generally been orientated away from gardens since the turn of the twentieth century. There are many plausible reasons for this state of affairs. First, many contemporary gardens exhibit what might be called traditional aesthetic values, such as beauty, elegance, and prettiness. Such aesthetic values became increasingly unfashionable during the century characterized by the philosopher Richard Shusterman as representing "The End of Aesthetic Experience."[45] Consequently, influential members of the

44 M. Conan, "Introduction: In Defiance of the Institutional Art World," in *Contemporary Garden Aesthetics, Creations and Interpretations*, ed. M. Conan (Washington, DC: Harvard University Press, 2007), 3.

45 R Shusterman, "The End of Aesthetic Experience," *The Journal of Aesthetics and Art Criticism* 55, no. 1 (1997), 29–41.

avant-garde artworld ignored gardens. Second, gardens have traditionally, though not always, possessed a high degree of sensuous content and a lesser degree of intellectual content. In a century during which the dominant, avant-garde artworld tended to focus on certain celebrated examples of (non-garden) art that possessed almost only intellectual content, and little or no sensuous content, gardens were again neglected. Third, gardens are not amenable to being transported, exhibited in galleries, or reproduced in facsimiles. They are, therefore, easily ignored by members of that artworld, whose preferred objects of interest and research do not have such drawbacks. Fourth, at most times in their history, art gardens have been expensive to make and to maintain at a level commensurate with their being works of art. Art gardens of any size have therefore commonly been created by and for wealthy individuals. This perceived elitism surrounding art gardens has made them unattractive to some members of an artworld, for whom the arts are understood to be a political tool to pursue egalitarian ideals, right social wrongs, or empower minority groups. Five, gardens cannot easily express anger, bitterness, or any form of negation. Until most recently, they have not usually been ironic or self-critical and nor have they typically shocked or confronted viewers. These characteristics made gardens unlikely candidates for art status within the avant-garde artworld, where part of the mandate for contemporary art has been to shock, disturb, and even offend.[46] Six, by their very nature, gardens are difficult to delineate in space and time. They also have a high degree of mutability and take unfashionably long to mature. So, the project of pinning down a garden in order for a member of any artworld to present it for consideration as an artwork was fraught with a complexity and difficulty unique to the garden artform.

None of the reasons just given entails that a garden artworld, whether running in parallel or as a subset of Dickie's artworld, does not exist. What

46 In a similar vein, landscape architects Dieter Kienast and Günther Vogt write that "the garden as a paradisiacal refuge is an elemental myth and by definition is thus a place of tradition and therefore not particularly suited as a medium for avant-garde experimentation." Dieter Kienast and Günther Vogt, "Die Form, Der Inhalt Und Die Zeit = Form, Content and Time," *Topos: European landscape magazine*, no. 2 (1993), 11.

they do amount to is an acknowledgement that the avant-garde artworld, and the philosophers associated with it, remained largely uninterested in gardens as an artkind.

Miller switches frequently between the terms "definition" and "theory," and uses them as if they are interchangeable.[47] Davies makes it clear, however, that, in the field of aesthetics, a distinction should be observed between "definition" and "theory."[48] He writes that "[a] successful definition [of art] must specify a set of properties all and only artworks possess and in virtue of which they are artworks," whereas "a theory of art can be more general in discussing what is typical or normative for works of art," and "is bound to reflect on art's significance within human lives and affairs."[49] By using the two terms interchangeably, Miller fails to distinguish between the characteristics of a procedural definition, such as Dickie's, and a functional theory, such as Beardsley's.

In the case of a procedural definition, the status of art is conferred by way of the attitudes and actions of a person or persons acting on behalf of the artworld. By contrast, in the case of Beardsley's functional theory of art, an object or event is a work of art if it possesses certain characteristics which give rise to aesthetic experiences in the viewer. The fact that a garden possesses these aesthetic experiences generating characteristics is what matters. Whether or not members of the artworld treat such a garden as a work of art is irrelevant to its art status.

47 For example, on page 70, Dickie's definition of art is referred to as a theory and a definition within the same sentence and, on page 71, Beardsley's aesthetic theory is referred to as a definition and, in tandem with the institutional definition, as a theory also. Miller is not alone among philosophers of aesthetics in using "definition" and "theory" synonymously. I draw attention to this aspect of her writing only because it has lead her to reach conclusions which, had a distinction between the terms been observed, would not have been possible.

48 Stephen Davies, "Essential Distinctions for Art Theorists," *Philosophical Perspectives on Art* (Oxford: Oxford University Press, 2010), 23–38.

49 Ibid., 32. I accept that Dickie's procedural definition of art is often referred to as the Institutional Theory, and I sometimes refer to it as the Relational Theory, but the fact remains that it is, in Davies' view, a definition, and not a theory.

In failing to distinguish between the meaning of the two terms, Miller ends up using Beardsley's theory to do the same discriminatory work as Dickie's definition. When she claims that Beardsley's theory/definition is inadequate because many people in the contemporary United States do not think of gardens as art, she is mistaking the condition required by the procedural definition with those required by the functional one, and her conclusion that gardens "do not fit our definitions of art" is therefore incorrect.[50]

Ross's and Miller's studies have each made substantial contributions to our philosophical understanding of art gardens. Their authors have been faced with the difficult task of attempting to accommodate gardens within *pan-art* conceptions, theories, and definitions of art. In the next part, I follow an alternative approach, gradually developing instead an artform specific account of gardens.

50 Miller, *The Garden as an Art*, 178.

CHAPTER 5

Enjoying Time(s) in the Garden

Time and Temporality in Philosophical Aesthetics

In 1991, two distinguished philosophers, Jerrold Levinson and Philip Alperson, published a comprehensive account of temporality in the arts entitled "What Is a Temporal Art?"[1] The authors answer the title's question by proposing a taxonomy of 13 individual conditions and one overarching condition, the possession of one or more of which is sufficient for an artwork's being classified as temporal.[2] They situate their argument in the context of well-known claims by Lessing, Zukerkandl, and others that some arts, for example music, are temporal and others, for example painting, are not, and they aim "to cover all that might conceivably be meant in predicating temporality of an art form."[3] They are interested in the possession and expression of time and temporality by standard or paradigm works in the genres of an art form, and they wish "to make sense of familiar intuitions about the arts."[4] In earlier chapters, I have been positing the idea that gardens are a temporal art. How, then, will gardens fare according to Levinson and Alperson's conditions?

1 Levinson and Alperson, "What Is a Temporal Art?."

2 Their conditions are presented as sufficient but it can be argued that some of them, including conditions 1, 2, and 3, are indeed necessary for an artwork's being temporal.

3 Levinson and Alperson, "What Is a Temporal Art?," 447.

4 Ibid., 441.

I have divided the authors' 13 conditions into two groups. Group One contains conditions 1–3. Had the authors chosen to consider gardens, these conditions could have uncontroversially accommodated them because the conditions are not in conflict with what might be termed the "received," though inadequate, account of gardens. Group Two contains conditions 4–13. If applied to gardens, these conditions either illuminate some special, though generally unremarked on, temporal aspects of them, or, when a condition cannot sensibly be applied to gardens, illuminate, by default as it were, some special temporal characteristic of gardens.

Conditions – Group One

Condition 1: Objects of the art form require time for their proper aesthetic appreciation or comprehension.

I agree with the authors that all art objects and events, and our experiences of them, have a durational aspect and that "this temporal aspect is ... not likely to be what anyone has in mind in thinking of the temporal arts as a special group."[5] It is incontestable that the "proper aesthetic appreciation or comprehension" of gardens takes time, for the simple reason that, except in the case of some few, extremely small gardens, a garden visitor needs to move about to experience gardens, and this necessarily takes time.

The condition does usefully describe one non-trivially temporal aspect of what could be considered a "proper aesthetic appreciation or comprehension" of gardens. T. S. Eliot describes how when composing a new work the creative artist responds to and is correctly influenced by her awareness of the implications of the unique temporal location of her new work within her own oeuvre and within its own creative tradition and chronology.[6] Similarly, Eliot argues, an "accurate" experience of *any* given work requires attention not only to any temporal relationships within that work but also

5 Ibid.

6 T. S. Eliot, "Tradition and the Individual Talent (1919)," in *Selected Prose of T. S. Eliot*, ed. Frank Kermode (London: Faber and Faber, 1975).

attention to the temporal relationships between that work and its precursors and successors in the genre.[7]

Because of gardens' characteristic mutability, Eliot's claim regarding the experience of a work are relevant to gardens. For instance, a traditional herbaceous border garden in a cold climate will have no plants at all visible in winter but in spring or summer it will be full of leaves and flowers. Many gardens seen once and then not again for, say, 10 years, will have grown and/or matured or decayed in many ways. It is fair to say in cases like these that a "proper aesthetic appreciation or comprehension" of the garden appropriately involves a knowledge of what that garden was like in winter, or 10 years ago, as well as a knowledge of what it may be, or be again, at some time in the future.

Condition 2: Objects of the art form require a significant interval of time for the mere perception or apprehension of their full extent.

Gardens are probably the candidates par excellence for being the most spatially extensive art form. Gardens like Le Nôtre's for the palace at Versailles, or the Australian Garden at Cranbourne, near Melbourne, make the highest durational demands – several hours – for the "mere perception or apprehension of their full extent." Such gardens exceed in this regard the demands of, say, the cathedrals of Chartres and Notre Dame, which are noted elsewhere in the authors' paper for their "extraordinary size and scale."[8]

It is uncertain how a viewer knows when they have experienced the "full extent" of a garden, or a building, and the question arises as to which criteria with respect to duration, angles, direction, time of day, seasons, etc., may be applied in answering this question. By comparison, it is relatively straightforward to know when one has experienced the "full extent" of, say, a 200-page-long novel, a 20-minute-long sonata, or a 300 × 450 mm painting.

7 A similar account of the relationship between a work and an artist's oeuvre is developed by Levinson. See: Jerrold Levinson, "Work and Oeuvre," in *The Pleasures of Aesthetics: Philosophical Essays* (Ithaca, NY: Cornell University Press, 1996).

8 Levinson and Alperson, "What Is a Temporal Art?," 446.

Condition 3: Objects of the art form require time in presentation, that is, they require performance or exposition of some sort over an interval of time; the parts of the artwork are not all available at any one moment, but only consecutively.

Gardens have a greater potential than any other art form to exhibit noticeable, non-aleatory changes while still retaining their ontological identity: bare trees break into leaf and blossom in spring; trees grow from sapling to forest giants; plants flower and then die. Gardens, however, are not performances, as the condition deems alternately necessary, because they do not have performers.[9] Gardens' most important constituents are living plants, and plants are not performers: sweet peas do not perform, they simply do what sweet peas do.[10] But the ways in which a garden designer arranges her garden elements result in certain temporal (and visual) events being exposed, juxtaposed, and counterpointed, and in this way the garden may be seen as an exposition, as alternately required by this condition.

Furthermore, a garden is never the same, it is always perceivably changing. So, in this additional sense, the time of the presentation of the garden is only limited by the garden's initial installation and final destruction, or its natural decay to the point where it is reasonable to say that the original garden no longer exists.[11] There is, if you like, one continuous "performance" while a garden exists.

Conditions – Group Two

Condition 4: Objects of the art form consist of elements or parts arranged in a linear order, with definite direction, from first to last.

9 For a thought-provoking presentation of an opposing view see: Salwa, "The Garden as a Performance."

10 See 167 for further discussion of this point.

11 Similar claims can be made for installations such as Damien Hurst's *A Thousand Years* (1990) or his *Let's Eat Outdoors Today* (1990–1), which feature living and decomposing elements such as items of food, maggots, flies, blood, and a cow's head. Such works require time for their presentation and not all aspects and stages of the component elements are available for viewing at any one time.

This condition does not accommodate most gardens, although there are some potential exceptions, such as the eighteenth-century circuit gardens.[12] However, the circuit gardens still have much to offer when they are experienced "backwards" or in some random order, especially when compared to say music, novels, and poetry, which result in noise or non-sense when experienced in those ways.[13]

Except for the very smallest gardens, the parts of most gardens can be experienced in any sequence. In this way, gardens are like the ludic novels of the twentieth century, which have a linear order, which start and finish, but which do not have an invariable sequence of parts or sections within their overarching structure. In these novels, as in gardens, the reader, or visitor, organizes the structure of the narrative, or visit, for herself.

There is one important way in which gardens differ from other arts that can be divided into sections. In gardens, and to a lesser degree in architecture, the direction, duration, and the experience of *moving between* sections of a garden are important contributors to the overall experience of a garden. Hunt notes that the transitional garden zones that give rise to such experiences are "the very essence of garden spaces."[14]

Condition 5: Objects of the art form are properly experienced in the order in which their elements are determinately arranged, and at a rate defined by, or inherent in, the artwork itself or its prescribed mode of presentation or performance.

Levinson and Alperson here write of the proper experience of an artwork at "a rate defined by, or inherent in, the artwork itself." This is not quite the case for gardens because in gardens the rate is inherent not in the artwork but in the particular natural materials which the garden maker chooses to

12 See Hunt's presentation of differing accounts of movement around Stourhead, and their importance and relevance to an adequate experience of that garden: Hunt, "Stourhead Revisited & the Pursuit of Meaning in Gardens," 330, 32.

13 See Cox's account of the positive, "non-iconographical" experiences of eighteenth-century visitors to Sourhead: Cox, "A Mistaken Iconography? Eighteenth-Century Visitor Accounts of Stourhead."

14 John Dixon Hunt, *Greater Perfections: The Practice of Garden Theory* (Philadelphia: University of Pennsylvania Press, 2000), 131.

employ and combine in her garden. Such materials "perform" at generally the same rate whether they are in a garden or not. Spring flowers always appear in spring and not in other seasons, trees grow at a rate determined by their biological nature and in response to the environment, and water springs from fountains for much of the year but not when frozen in winter. Therefore, it may be said that although gardens are not performances, they are presentations, the rate and order of whose elements is determinately arranged in what might be understood as a joint venture between nature, the garden designer, and the gardener.[15]

Condition 6: Objects of the art form are such that non-temporally extended parts of the object do not count as aesthetically significant units of it. That is to say, such parts are not isolatable for study in a way that contributes significantly to the full experience of the object.

Although I am arguing that gardens are indeed a temporal art, there is an exhaustive literature and commentary in other formats with respect to what might be termed "frozen moments" in gardens. In fact, because of the limits of language and photography, the majority of descriptive garden writing and images necessarily involves the description of static scenes or series of scenes, and, despite their obvious limitations, it is counter-intuitive to say that such scenes do not "count as aesthetically significant units" of a garden, nor that "such parts are not isolatable for study in a way that contributes significantly to the full experience" of a garden.[16]

15 Stephen Davies claims that in some architecture and gardens, the designers of both can set out to deliberately influence the rate and direction of a visitor's movement. (See: Davies, *Philosophical Perspectives on Art*, 139.) This is indeed a theoretical possibility. However, I claim that, apart from some ritualistic uses, it is seldom the case that garden visitors are influenced in this way. (See 124–25.)

16 References to the ekphrastic challenges faced by those who present accounts of gardens are well known. See, for example: Birksted, "Landscape History and Theory: From Subject Matter to Analytic Tool," 8; John Dixon Hunt, "Beyond Ekphrasis, Beyond Sight, Beyond Words. …" <http://www.doaks.org/research/garden-landscape/scholarly-activities/past/sound-and-scent-in-the-garden/abstracts>; *Greater Perfections: The Practice of Garden Theory*, 131; Salwa, "The Garden as a Performance," 385.

Condition 7: Objects of the art form are about time, or our experience thereof, in some significant way.

If the preposition "about" in this condition is taken to mean something like "having as its subject," or "concerning," then the condition excludes gardens. I have earlier questioned Ross's accounts of gardens' ability to function in the ways poems and paintings do, and therefore their ability to represent or be discursively about something.[17] Gardens, although inescapably temporal, are not usually *about* time. Given that the time of a garden is the same as the time outside a garden, one could say that gardens are *in* time.

Another possibility is that the preposition "about" references Danto's notion of "aboutness," that is, the quality an object or event has that distinguishes artworks from mere real things.[18] In this case, gardens turn out to be a good exemplar for Danto's claims because their materials, and the rates at which they change and we experience them, are generally the same both inside and outside the garden. And in this way gardens are inevitably "about" time.

Condition 8: Objects of the art form use time as a material, or as an important structural feature.

There are two important ways in which gardens comply with this condition. First, gardens, and in particular their natural materials, necessarily exist in, rely on, and have their effects in time. Second, the *experience* of gardens is importantly temporal because almost all gardens require a visitor to move around, at whatever speed and in whatever direction, in order to experience the garden.[19] Usually such movement is not "choreographed" by the garden maker. Such movement, and a visitor's own choreography of it, take an open-ended amount of time (and space).

17 See 82–5.

18 See: Arthur C. Danto, *The Transfiguration of the Commonplace: A Philosophy of Art* (Cambridge, MA and London: Harvard University Press, 1981), 113.

19 Birksted reminds us that a garden beholder is perpetually "mobile … in gardens' three-dimensional space and time" (Birksted, "Landscape History and Theory: From Subject Matter to Analytic Tool," 6).

Condition 9: Objects of the art form generate a kind of time that is peculiar to them, that exists for a perceiver only in and through experience of the work.

Gardens do not generate a peculiar kind of time. Instead, their time is the time of the non-garden world, and the time of our garden experience is also the time of our experiences outside the garden. Flowers open and ice creams are consumed at the same rate whether in gardens or not.

Condition 10: Objects of the art form represent a series of events in time distinct from the series of events constituting the art object.

This condition does not accommodate to gardens for reasons already given: gardens cannot generally *represent* events; and, even if they could, gardens could not do so in a "time distinct from the series of events constituting the art object."

Condition 11: Objects of the art form are created in the act of presentation, so that the time of creation, time of presentation and (usually) time of reception all coincide.

Condition 12: Objects of the art form require presentation in a time lived through and by the presenters.

These two conditions are particularly relevant to the account of the experience of gardens that I present in the Chapter 8. There, I argue that the actual, physical garden is akin to a pre-existing work which one or more improvisers, or garden visitors, uses as the basis for an improvisation. Although the seminal work is important and necessary, and perhaps a work of art in its own right, in such cases it is the improvisation – the garden *experience* – that is the pertinent and valuable "work of art."[20]

Condition 13: Objects of the art form lack relatively fixed identities over time, but are rather mutable and shifting.

20 My account of the *experience* of gardens at 172–79 is importantly different from Salwa's account of gardens as performances, which account would, however, be neatly accommodated by these two conditions. (See: Salwa, "The Garden as a Performance.")

Whether some gardens change more than the folk arts and arts with an oral tradition, which the authors cite as examples of arts accommodated by this condition, is debatable. Miller has claimed that gardens do have a fixed identity and that it is related to their geographical location.[21] However, some show gardens have been relocated. A garden may also be neglected to the point of its becoming a non-garden, or it may be destroyed in some way, yet still that new, non-garden object retains its same geographical identity. So, the identity of a garden depends on more than just its geographical location: a garden is not just a matter of *where* it is, but of *what* is where it is. And, in the case of some garden styles, our conception of that "what" needs to be fluid enough to accept gardens' "mutable and shifting" identities.

Considered against the criteria of Levinson and Alperson's taxonomy, gardens are clearly temporal works of art. Gardens satisfy nine of their conditions, the satisfaction of any one of which the authors regard as sufficient for an artwork's being temporal. They also show up the need for an additional condition to be added to the taxonomy. This new condition acknowledges characteristics of fundamental importance in gardens, in much environmental and land art, and, in part, in some installations and architecture.[22] It reads: "Objects of the art form are aesthetically dependent to varying degrees on the transitions, movements, actions, and patterns of biological, diurnal, seasonal, climatic, and sometimes geological changes, most of which occur in temporally experienced sequences."

The authors allow that their criteria will "permit 'temporal art' to have application to art forms which do not currently exist, but which are theoretically possible," so it is therefore tempting, if mischievous, to speculate that if gardens were to be invented in the future the authors might classify them as "temporal art."[23] There are two possible reasons to explain gardens' omission from Levinson and Alperson's category of temporal art. First, from the early nineteenth century onwards, gardens have typically been ignored by philosophers of art, a situation which is only now slowly beginning to

21 See Miller, *The Garden as an Art*, 76–7.

22 In proposing this new condition, I owe a debt to Mara Miller for her writings on gardens and time.

23 Levinson and Alperson, "What Is a Temporal Art?," 441.

change.[24] Second, until recently, philosophers of art, and the conceptions, definitions, and theories of art they have espoused, have disregarded or even disdained the possibility of unruly, changeable natural objects (plants), and the natural environment generally, being acceptable components of art.[25] In particular, accounts of the ontology and preferred mode(s) of experience of art have been inimical to the reality of gardens and the garden experience.

Time and Temporality in Gardens

I begin with a caveat. Gardens are temporal arts but they are not *only* temporal arts. Gardens' aesthetic appeal and value derive from their being both temporal and pictorial (spatial) arts. They are a four-dimensional art, and they share this bi-(multi-)modality with some other arts.[26] Discussing the contribution of temporality to the overall aesthetic value of a garden is therefore not a straightforward matter and I do not mean to imply that it is possible to consider or quantify altogether successfully the visual and temporal aspects of gardens as if they were discrete, independent aspects of a work. Nonetheless, investigating temporality on its own remains a worthwhile task, and a necessary one is we are to understand fully the processual, and iterative qualities of gardens.

A merely descriptive listing of types of change in a garden would be of limited value if those changes were not themselves resonant of temporality. It is not mere changes, but the meaningfulness and values associated with

24 For Cooper's account of the lack of interest that twentieth-century philosophy, and some other disciplines, have shown towards the garden see: Cooper, "Foreword," x.

25 According to Hunt, the root cause of the majority of "problems" gardens have caused for philosophical aesthetics stems from the discipline's unwillingness to engage with the reality of gardens' "actual messy, material, and changeful world." See: Hunt, *Greater Perfections: The Practice of Garden Theory*, 7.

26 Birksted describes a garden existence, and a garden beholder's experiences in it, as occurring in "three-dimensional space *and* time" (my emphasis). See: Birksted, "Landscape History and Theory: From Subject Matter to Analytic Tool," 6.

them – their temporality – that is of interest. Temporality is different from time and may be characterized as our response to the effects of the passage of time on objects, including ourselves, and especially on objects whose existence is importantly temporal.[27] Temporality is the condition or quality we are aware of a thing's possessing by virtue of its being in time or associated with time. Temporality is not something that exists independently; rather, it is a product of our experience of objects and events.

The temporality of an object is linked to our perceptions of the qualities the object possesses with respect to its duration, persistence, and ephemerality, its physical stability and manner and rate of change(s), its temporal "location," or order in which it occurs, and so on. We perceive changes occurring in objects and events, or we are aware of the potential for such changes to occur, or not occur, and we understand the changes, or absence of changes, to have a significance for the object ranging from trivial to, quite literally, vital. We assess such changes for the importance, interest, and value they may have for us and for the object in question. Our assessments may be based on the significance of the changes, or lack of changes, with respect to an object's financial value, safety, viability, or physical existence, or with respect to any symbolic and aesthetic considerations for the viewer of changes, or lack of changes, that the object undergoes or has the potential to undergo.[28]

How Gardens Change

Changes that occur in gardens are the outcome of inter-related processes including, principally, the actions of skilled, knowledgeable gardeners, who manage and maintain gardens' materials in the expectation of desired

27 A dictionary definition of "temporality" reads, in part: "The quality or condition of being temporal or temporary; temporariness; relation to time." (See: "Temporality, N.," September 2016 edn, *OED Online* (Oxford University Press), <http://www.oed.com/view/Entry/198948?redirectedFrom=temporality>).

28 For an account of rhythmical ("musical") aspects of the experience of temporal events in gardens see: Barwell and Powell, "Gardens, Music, and Time."

outcomes, the effects of weather, diurnity, and seasons, which result in both predictable and unexpected outcomes, and the natural growth and decay patterns that are characteristic of all living things.

Changes resulting from these processes can be grouped into broad categories related to maturation, maintenance, biological change, cyclical change, occasional change, and planned change. Changes in any one or all these categories are not always conspicuously present at any given time. Moreover, changes in some categories are perceivable only to a retuning visitor while changes in other categories are perceivable only within the timespan of a single visit, whether of a frequent or first-time visitor. Changes in all these categories offer potential reasons for our aesthetic interest in and valuing of the gardens possessing them.

Maturation

A designed garden might be described as finished when it is newly and completely installed. Alternatively, a designer might specify that her work will be finished at some time, say 20 years after installation, by which time structural tree plantings will be starting to develop their adult shape and create the spaces the designer intended; hedges will have grown into dense green walls; the 100 daffodil bulbs originally specified to be planted will have multiplied into a carpet of several thousands; and introduced streams and watercourses will be looking "natural."

The question of when a garden might be deemed to be finished, or complete, is one which betrays a misunderstanding about what gardens are. While there may be desirable, pseudo-permanent stages to reach during their existence, gardens are processes, or events, as much as they are unchanging objects. Gardens must change continuously in order to persist and, in this way, they are unlike works in any other traditional art genre.

However, leaving aside the case of short-lived gardens designed for temporary exhibitions, many garden designers do have in mind some idea of a preferred "finished product," and the journey towards that ideal state, and the extension of it once reached, are closely related to how a garden is maintained.

Maintenance

At first glance, the maintenance of a garden may seem to be a process analogous to that of conservation for painting and sculpture, but this is not so. Conservation involves returning an object to an aesthetically preferable state, often a work's original state, in which state it should ideally remain, unchanging. Maintenance of gardens does not involve a one-off, remedial action of this type. Instead, it is an essential, continuous intervention, which is rarely concerned with returning a garden, for whatever brief instant, to its original state, even if that state is determinable and desirable. It involves working collaboratively with the temporal processes of change that are continuous and inevitable in gardens. It involves an acknowledgement that among the traditional art genres, gardens manifest a unique temporal nature.

Because gardens and their component elements are largely living, maintenance usually endeavours to balance the inevitable tensions between the stability of the garden-object and the process of continuous change. In the case of some garden styles, such as Italian gardens from the Renaissance, maintenance does not set out so much to encourage or discourage changes brought about through temporal processes as much as to deny time any role at all in the garden, by resisting its effects. In such apparently timeless gardens, frequent and sometimes daily maintenance tasks are carried out by the gardeners ensure that no changes appear to have occurred.

The constant effort required to maintain gardens in whatever their desirable status quo may be has no equivalent in other traditional art genres and is further evidence of the unique temporal status of gardens. It reinforces that time and temporality play a significant role in a garden's physical existence and dependent aesthetic value and interest, and that, in gardens, object and event are inseparable.

Biological Changes

Plants, whose presence is characteristic, distinctive, and usually constitutive of gardens, are living organisms. Because they are alive, plants necessarily change continuously and in accordance with their own internal programmes. While the direction and rate of change can be influenced by human and other interventions, and while plants may be treated in ways that cause them to thrive or allow them to recover from illness, plants cannot be "instructed" to act in ways for which they are not internally pre-programmed. Plants, therefore, do not perform when they change, as actors and dancers do when they follow scripts and choreographies. Plants simply do what they must do. And in this way, gardens differ from the other traditional art genres that use living materials because gardens are not performances.

The biological changes that are distinctive of gardens offer one of the most important reasons for paying aesthetic attention to and deriving aesthetic pleasure from gardens. These biological changes can be grouped into four categories, while allowing that the boundaries between the categories are somewhat permeable.

The first category of biological change is concerned with the maturation of the garden as a whole. The designer and gardeners invest much energy in guiding a garden towards its mature state and preventing its decline into senescence. For example, they carefully calculate the relative speeds of growth and the flowering times of different plants to ensure that a (southern-hemisphere) herbaceous border will be "at its best" in, say, the month of January. They do this because it is thought that a particular stage, or stages, in the life of the garden are most worthy of aesthetic attention and will afford maximum aesthetic pleasure to viewers of the garden. In behaving in these ways, designers and gardeners acknowledge that gardens are in large part biological processes which, while they may be influenced, operate in accordance with the biological programmes of their constituent plants.

The second category concerns the natural processes of individual plants. These processes are to be distinguished from the products of such processes, which I discuss below. When I refer to processes I am referring to activities such as growing and shrinking, living and dying, furling and unfurling, opening and closing, flowering, fruiting, seeding, and so on.

Some of these processes may be perceivable by the casual visitor, but more often they reveal themselves to the dedicated gardener or frequent visitor. However, it is possible for a casual visitor to be imaginatively aware that the apparently static moment she is observing is a step in a continuous sequence of, say, growth. All these, and similar processes are constitutive of gardens and aesthetic attention is properly directed towards them.

The third category of biological change is concerned with the products of processes of change. In one sense, the distinction I make here between process and product is false because the fruits, flowers, and so on are themselves processes as much as they are products. However, the distinction is reasonable because of the importance in gardens of flowers and fruit *qua* products, not processes, suited to aesthetic attention and generative of aesthetic value. Some gardens exist primarily to display their products, traditional rose gardens being a well-known example. In such gardens, aesthetic attention to a flower- or fruit-product is similar to the aesthetic attention paid to an unchanging painting or sculpture. But it is different from attention paid to a painting or sculpture because it is attention to a non-permanent object, which cannot be restored once it deteriorates, and because it is a multisensory attention, potentially engaging not only sight but also taste, smell, and touch.

Cyclical Change

Diurnal cycles are regular. Unlike seasons, where the effects of spring may come "early" one year and "late" the next, day follows night in an entirely predictable though varying diurnal and annual pattern. A garden's appearance changes depending on whether it is viewed by day or night. These changes are caused by the presence or absence of light and therefore they affect primarily how a garden looks and, consequently, its colours and shapes, apparent size, views, mood, and aspect, and so on. Some garden plants react quite conspicuously to the presence and absence of light. For example, many plants close their flower petals at night, some plants flower only at night, some plants release scents at night, and some plants close their leaves at night. The flower heads of other plants react to gradual changes in the sun's position by tracking its movements throughout the day.

The progression of the seasonal cycle is responsible for some of the most conspicuous changes in gardens. Bare trees in winter, new scents in spring, prolific flowers in summer, and brightly coloured leaves in autumn are all familiar garden "pictures." The conspicuousness of the changes resulting from this progression varies and is dependent on the geographical location of a garden, the garden's style, and the garden's plant communities. Thus, a formally devised, tropical garden at an equatorial latitude will change little from season to season compared to an informally conceived woodland garden or mixed flower garden in a temperate climate. Temperate climate gardens are often designed to be "at their peak" during a particular season, or seasons, and the differences between gardens of this sort "at their peak" and at different times of the year can be extreme, and are unparalleled in other traditional art genres.

Occasional Changes

Occasional changes comprise a group of changes that significantly affects all gardens. Principal among these are changes attributable to weather, two powerful agents of which are wind and rain. Occasional wind may rustle leaves and ripple water, or it may be sufficiently strong to make walking difficult and even to fell trees. Constant wind may alter trees' profiles and generally stunt plants' growth. Rain, by its presence, may encourage growth and cause colours to glow in its aftermath. It may also make garden visiting unpleasant and it may cause floods. By its prolonged absence, rain may cause whole gardens to die.

No artworks in any other traditional art genre are affected by occasional changes to this degree and even if they are affected, as is potentially the case with outdoor sculpture and outdoor performances, the changes wrought are usually neither constitutively important to the artwork nor are they generally appropriate targets of aesthetic attention on the part of informed viewers.

Planned Changes

There are two ways in which planned changes may affect gardens and a visitor's experiences of them. The first type of changes has parallels in the performance arts. In this way, fountains and lighting systems may be activated from time to time in gardens because the designer, owner, or gardener intends this to happen, just as stage effects – for example the visual and sonic components of a storm – can be activated when required by the stage directions or director during the performance of a play. In the case of gardens, these changes are not distinctive, nor do they provide grounds for supporting the conclusion that gardens relationship with the passage of time is unique or constitutive of them.

The second type of changes is constitutive of gardens and is unique to them among the traditional art genres. These changes concern the installation, removal, or modification of plants in accordance with a schedule determined by the designer, gardener, or owner. These changes are similar to some of the maintenance tasks described earlier. They include activities such as the planting out of annuals, the removal of sensitive plants to orangeries in winter, the planting of shade-loving species under trees when young trees have become sufficiently umbrageous, and the thinning of initially over-planted areas as trees mature. While these activities may be categorized as simple maintenance, they also contribute to the status of gardens as a unique art form because they exemplify the ways in which gardens are designed not only to persist but also to be regularly changed in specific ways that are aesthetically important, in accordance with a designer's or owner's instructions. Gardens, uniquely, are art objects that remain themselves while continuing to change and be changed in significant ways that are aesthetically interesting and valuable.

In summary, gardens are (a) spatio-temporal works, that (b) are not performances, but (c) whose elements are continuously changing at a variety of rates and in different ways, and (d) whose components include living materials, and (e) whose components can sometimes be removed and/or replaced, and (f) whose aspect can be radically yet acceptably changed owing to the continuous influences of seasonal, climatic, and diurnal changes. Some of these characteristics have been, until recently, unique to the artform of gardens. All these characteristics provide reasons for our aesthetic interest in and valuing of gardens. They also all affect the ontology of gardens, which is the subject of the following chapter.

CHAPTER 6

What *Sort* of Artwork Are Gardens?

Gardens meeting certain criteria are considered to be works of art. It is therefore reasonable to expect ontologies making claims to adequacy in the case of artworks to be able to accommodate them. Principal among the issues an ontology adequate for art gardens needs to deal with are gardens' sometimes extreme mutability and their status as singular, not multiple, works of art. None of the art ontologies presented in this chapter has been developed with gardens specifically in mind but examining them from that point of view is an interesting and informative exercise.

Concerning the variety and complexity of ontological positions currently on offer, David Davies suggests that "in matters of ontology, it might be thought, contemporary philosophy of art has almost no shame!"[1] However, Davies' provocative sentiment aside, it is possible to identify common issues and tasks which an ontology of art needs to address. Amie Thomasson and Stephen Davies are two well-known contemporary writers on the ontology of art.[2] Although they differ radically in their *approaches* to addressing art ontology's important tasks they do share a broad view of what those important tasks are. Their key questions may be summarized as: what sort of objects are works of art? How do they exist and persist? How do we identify them? What category do they belong to, and what further information might we derive from their categorization? [3]

1 David Davies, *Art as Performance* (Malden, MA and Oxford: Blackwell Publishing, 2004), 127.

2 For ease of reading, in the remainder of this chapter I refer to David Davies as "D. Davies," and Stephen Davies as "Davies."

3 Stephen Davies, "Ontology of Art," ed. Jerrold Levinson, *The Oxford Handbook of Aesthetics* (New York: Oxford University Press, 2009), <http://www.oxford-handbooks.com/view/10.1093/oxfordhb/.../oxfordhb-9780199279456>. § 1; Amie

To examine these questions in the case of gardens, I have selected ontologies developed by Davies, Wollheim, and Currie, all of whom are or have been (Wollheim died in 2003) prominent in this field. In brief, Davies supports the view that works of art are in some way related to the physical objects constituting them, Wollheim supports, but later modifies, the view that artworks (aesthetic objects) bear an identity relationship to the physical object constituting them, and Currie supports the view that works of art are the results of actions. My interest in their ontologies is limited to pointing out the ways in which they are, or are not, adequate to the case of gardens. A full analysis of them would take a book of its own!

Gardens Are Physical Objects

In *The Philosophy of Art*, Davies defends a position he describes as "ontological contextualism."[4] Many aspects of this position are uncontroversial and, ontologically speaking, mainstream because they are shared to varying degrees and with varying emphases by many other contemporary ontologists for whom an artwork is identified with or is in some way related to a specific physical object or event. Davies asserts that artworks are public items – objects or events – and that they are created rather than discovered. He asserts that their designed form and content are inseparable from their medium and that whether they are singular or multiple in their instances depends on which genre they belong to. If they are multiple they exist in one of three ways: they can either be presented as a performance, they can be specified in encoded forms, such as on a digital file or woodblock, or they can take the form of an exemplar, such as a novel or poem, produced by the artist at the time of the work's creation. Finally, he argues that, generally speaking and his contextualist views notwithstanding, artworks'

L. Thomasson, "Ontological Innovation in Art," *The Journal of Aesthetics and Art Criticism* 68, no. 2 (2010).

4 Stephen Davies, *The Philosophy of Art* (Malden, MA: Blackwell, 2006), 81–9.

identities are fixed at the time of creation and that "factors crucial to the work's identity [do not] alter after its initial creation."[5]

Davies makes three claims pertinent to the ontology of gardens. In answer to his question, "does an artwork have an evolving identity, much as a living person does?", he makes his first claim: "Some may do so. Gardens, supposing some qualify as artworks, are intended to change with the growth of the flora and the yearly cycle. These works change while retaining their basic identity only because their media and constituent elements are living. It is less plausible to suggest the same for other artworks."[6] Davies second claim is related to his first one but appears to partly contradict it. He writes: "For art, we can account for the manner of its existence without resorting to the special kind of ontology that is used to explain the persistence through change of the identity of living creatures."[7] Davies' third claim is that some architecture shares with gardens a temporal nature because the designers of both can set out to deliberately influence the rate and direction of a visitor's movement around a building or garden. In such cases, he writes, buildings and gardens are less like "symphonies, films and drama than like novels."[8] He goes on to note that although a fixed view, say from a particular window, of a garden through the seasonal cycle may constitute a temporal experience more akin to "a symphony or drama than a novel," he is not aware of any building affording a similar experience, even though he concedes that a site-specific building could potentially offer such an experience of time's passage.

Davies' first claim, concerning the acceptability of change in gardens, is supported by his statement that such change is acceptable "only because

5 Ibid., 97.

6 Ibid., 96. In this section of his text, Davies has been referring to Margolis's view of the evolving identity of artworks being constituted in part by socio-historical factors, including interpretation. Davies rejects this view; but the changes in gardens' identities he refers to here are not of the type Margolis's conception envisaged. For Margolis's exposition of his own theory, see: Joseph Margolis, "What, after All, Is a Work of Art?," in *What, after All, Is a Work of Art?: Lectures in the Philosophy of Art* (University Park, PA: Penn State Press, 1999).

7 Davies, *The Philosophy of Art*, 98.

8 *Philosophical Perspectives on Art*, 139.

their [i.e. gardens'] media and constituent elements are living."[9] At this point in his text, Davies is countering the claim of the American philosopher Joseph Margolis that a "work's context continues to affect its identity after its creation."[10] Davies admits that the identity of gardens may change after their creation but that such change occurs, and is acceptable, only because gardens contain living elements which perforce change, and such change does not constitute proof of Margolis's claim. Although Davies had other aims in mind when making this point, it is a welcome statement regarding gardens' ontology. However, it is important to acknowledge that significant changes occur in gardens for reasons other than plants' mere aliveness. For instance, plants may change, or be added, removed, or altered as a result of the effects of climate or disease, or because of changes in ownership or fashion, and such changes have the potential to affect the garden's identity.

Furthermore, it is not the case that all gardens or all gardens' "media and constituent elements are living."[11] Some celebrated gardens are very sparing in their use of non-living materials. Many well-known classical Spanish, Italian, and French gardens rely for their effects as much on statuary, paving, walls, terraces, urns, steps, and water as they do on plants. When such gardens do rely on plants, those plants are often manicured into lawns, hedges, and topiary and are thereby presented as seemingly unchanging (that is, "non-living") garden elements. Gardens, then, are complex entities. They contain significant living and non-living, changing and non-changing elements, on all of which, and on the interactions between which, gardens' aesthetic properties supervene.

Davies second claim is that art ontology has no need to resort to "the special kind of ontology that is used to explain the persistence through change of the identity of living creatures."[12] The special kind of ontology to which Davies refers is one that allows us to recognize a person as being

9 *The Philosophy of Art*, 96.

10 Ibid.

11 "Ontology of Art," 96. To be fair, Davies is not setting out to deny that gardens may contain non-living objects. His concern is with rejecting Margolis's claim, and his rejection stands whether gardens contain non-living materials or not.

12 *The Philosophy of Art*, 98.

the same person even though she may be blond and thin at one point in her life and bald and obese and even "he" at another. Such an ontology is very similar, though not identical to the ontology required to account for the identity of a garden as it changes from, for example, being newly installed in winter to being mature 30 years later in summer.[13] Furthermore, the very changes which Davies is probably implying when he talks of "the persistence through change of the identity of living creatures" are themselves at the heart of gardens. Other arts, such as the performance arts of dance, drama, and music, involve changing, living creatures as a constituent part of their existence but aesthetic interest does not usually supervene on actors', dancers', or musicians' actual livingness. Although bodily changes in dancers, such as those that result in a *jeté* or an *entrechat*, are constituent of the performance, other simultaneous physiological changes in the same dancers, such as their growing (imperceptibly) older or having a cold, are not. In the case of gardens, it is precisely physiological changes such as ageing, senescence, growth, flowering, and so on that are at the core of the existence and process of gardens, and I believe that Davies has allowed for this to be the case in his first claim.

It is not clear how, or if, Davies intends to make a distinction between the phrase, "living creatures," cited in the paragraph above, and his clause, "media and constituent elements [that] are living," quoted earlier.[14] If he does not intend to make a distinction between these categories, then his claim that there is no need to invoke "the special kind of ontology that is used to explain the persistence through change of the identity of living creatures" is either mistaken or, alternatively, Davies has at this point decided to exclude gardens from the "art" category. If Davies does mean to imply a difference between "living things" and "living creatures," a clarification of what that difference entails would be useful. If it is simply a matter of distinguishing between (a) an ontology used to explain the persistence through change of the identity of living *creatures* and (b) an ontology used to explain the persistence through change of the identity of living *things*,

13 I return to this claim on p. 173.

14 Recall that this clause identifies components, the possession of which justifies an artwork's having "an evolving identity, much as a living person does."

then that is an appropriate, subtle distinction to make. However, I suspect that in both cases he is referring to biological events such as birth, growth, flourishing, and decay that are common to living "creatures" and living "elements," and through which processes those creatures' and elements' identities are sustained, and, if that is the case, his second claim is mistaken.

Also relevant to Davies' second claim is his statement that "qualities of the *medium* are likely to become part of the artwork's content and message" (my emphasis).[15] Ironically, he is making this statement as part of a discussion of avant-garde artworks which self-destruct, such as works involving dying butterflies or rotting meat, and not with reference to gardens, which characteristically grow and flourish. However, his statement does apply with equal aptness to gardens, and invites further reconsideration of his claim that art ontology has no need to resort to "the special kind of ontology that is used to explain the persistence through change of the identity of living creatures."

Davies third claim, that some architecture shares with gardens a temporal nature because the designers of both can influence visitors' movements and rate of progress, is right as far as it goes. Garden designers, and gardeners, certainly can and do manipulate the rate and direction of progress around a garden and they have available to them a wealth of techniques to achieve this. For example, at Stourhead, a visitor's progress around the grounds is influenced not only by the usual means of paths and bridges but also by a complex narrative and iconographic scheme. However, most gardens are equally temporal for the simple reason that they are not small enough to be taken in at a single glance. A visitor generally needs to move in space and time to experience all but the smallest gardens and, in this limited sense, gardens are indeed usually temporal artworks. However, such movement need not be influenced in any way by the design of the garden maker and may simply derive from a visitor's whim, special interest, timetable, or desire to walk in the shade. And, although Davies does not here mention it, the same sort of temporality is also part of the experience of

15 Davies, *The Philosophy of Art*, 87.

architecture, even in cases where movement is neither directed, foreseen, or even sanctioned by the designer.

Davies' third claim includes an interesting statement about how the version of temporality now under discussion – that is, the temporality that is generated by virtue of a visitor's choosing to be stationary and/or to move around a building or garden at differing speeds and in different directions – might compare to the temporalities generated in different art genres. He writes that when buildings (and presumably gardens) exhibit this type of temporality they are less like "symphonies, films and drama than like novels."[16] This comparison, however, does not pick up on an important temporal difference between novels and architecture (and gardens). In novels, although a reader may jump from anywhere to anywhere else in the text, and may even read the whole text backwards, this is not the way in which a novelist would normally expect her work to be experienced, and nor is it the way in which a novel's themes, plot, characterizations, and so on will make the best sense and be most aesthetically effective. But such is often not the case with architecture and gardens. Although a visitor's progress may sometimes be influenced by a designer's "programme," any *directed* progress through buildings or garden is much more likely to stem from ritualistic requirements, practical concerns such as crowd control or safety, or from commercial imperatives that ensure that your visit terminates in the souvenir shop.

In addition to the three claims already examined, Davies suggests, and then rejects, the interesting idea that architecture may usefully be considered a type of performance. It is an idea that is also interesting in the case of gardens, but I reject it for the same reasons, *mutatis mutandis*, as Davies.[17] Gardens, though, are not unchanging artworks, and the notion of performance is not altogether unhelpful in understanding what is going on in them. Performance works admit of a complex variety of interwoven spatial and temporal changes and they "encompass certain parameters of indefiniteness," and in these ways they have much in common with the

16 *Philosophical Perspectives on Art*. 139.

17 Ibid., 141–5. I examine the singularity/multiplicity and performance/non-performance features of garden artworks in detail at 162–72.

particular type of four-dimensionality that is characteristic of gardens.[18] Viewing gardens as a type of performance similar to, say, music may be useful and lead to fresh insights, and Davies himself suggests as much when he writes that a fixed view of a garden throughout the cycle of seasons may offer a temporal experience more akin to "a symphony or drama than a novel."[19]

Gardens too "encompass certain parameters of indefiniteness," even though they are not performances. Their indefiniteness invites clarification of Davies' overarching description of artworks as *public* objects or event. Many gardens are indefinite in their spatial extent and this is a significant, constitutive element of them. In Chapter 1, I referred to the permeable aural boundaries of Tupare, and many examples of permeable visual boundaries are to be found in gardens which intentionally or otherwise include "external" elements such as a view of a church steeple or the sea. These elements can be a central feature of or even provide a raison d'être for the gardens that possess them. If artworks are indeed public objects or events then, in the case of gardens, there must be flexibility with respect to the spatial extent of the art object. The garden-object's perceptual boundaries may extend significantly beyond a garden's cadastral or similar boundaries. In cases when there is no shared agreement on where a garden stops, the public object that is an art garden must be acknowledged to be spatially indefinite.

In summary, Davies' acknowledgement of gardens' inherent changefulness is welcome; his implied claim, that gardens' retention of their identity in spite of constant change is not plausibly accounted for by an ontology similar to that employed in the case of living creatures, is debatable; and his account of an artwork *qua* public object might be clarified in order to be able to accommodate artworks that are partly indeterminate spatially but whose aesthetic qualities are intimately related to their spatiality, like gardens.

18 *The Philosophy of Art*, 90.
19 *Philosophical Perspectives on Art*, 140.

Gardens Are Actions

In his *An Ontology of Art*, Currie presents and defends two hypotheses, the Action Type Hypothesis, and the Instance Multiplicity Hypothesis.[20] Currie does not test these hypotheses against the case of gardens. However, in arguing below for their inapplicability to gardens, important constitutive aspects of gardens are usefully illuminated.

Action Type Hypothesis

The Action Type Hypothesis claims that any artwork has two jointly necessary components: a perceivable structure – of words, sounds, visual elements, etc. – and a unique heuristic path. Currie links together a work's perceivable structure, say, a sonata, and the actions by which it was arrived at and/or executed, to form the single object we know as a work of art. This art object is neither created nor discovered, as others would have it; rather, it is *enacted* by the artist.

The first component – a perceivable structure – is unremarkable and is to be found, albeit with differing emphases and limits, in ontologies developed by Wollheim, Goodman, Levinson, and others. The second component – a unique heuristic path – is, while not unique to Currie, controversial in his presentation of it.[21] What Currie seeks to introduce into the concept "art" by using the term "heuristic path" is an acknowledgement of the constituent importance of the ways in which and the actions

20 Currie, *An Ontology of Art*.

21 An ontology in some ways similar has been developed by D. Davies. (See: Davies, *Art as Performance*.) While Currie argues that the perceivable structure of an artwork is discovered via a heuristic path, D. Davies argues that the perceivable structure of an artwork is created via a performance. My objection developed in this chapter with respect to the conflict between Currie's concept of an artist's discovery of a (completed) work and the reality of a changing garden can also be applied to D. Davies' notion of an artist's performance of a (completed) work.

and thoughts by which an artist forms a work's perceivable structure. He claims that artists who follow different heuristic paths but end up designing perceptibly indistinguishable structures have in fact created different artworks.

For example, if the Dadaist Duchamp were to have produced a perceivable structure identical to *Mona Lisa*, it would not be the same work of art as *Mona Lisa* because we can assume that in producing it Duchamp would have followed a different heuristic path from Leonardo's. Leonardo's heuristic path would possibly have included the requirement to earn a living by painting a commissioned portrait, a wish to reference paintings of the Madonna and, innovatively, the decision to employ a landscape scene as a background to the figure. The heuristic path enacted by Duchamp in producing his work would almost certainly have been governed by different matters, including a wish to express a sense of humour and a decision to be true to the tenets of Dadaism. Therefore, Leonardo's *Mona Lisa* and Duchamp's painting are completely different works of art even though their perceivable structures are physically indistinguishable.[22]

With respect to gardens, there are two problems with this hypothesis. First, while a particular perceivable garden structure *could* be the result of following more than one heuristic path, this must remain only a theoretical possibility if Currie intends that identical, enacted perceivable structures can exist simultaneously. Gardens are essentially site-specific works. Attempts to produce, via whichever heuristic path, the "same" perceivable garden structure on a different site, or, absurdly, to reproduce it on the site of an existing, identical perceivable garden structure, must fail. This is not the case for non-garden arts. For example, however unlikely it may seem, it is *actually* possible for an open-ended number of perceivable structures identical to what we know as *Mona Lisa* to be created via different heuristic paths and to co-exist. Furthermore, even if the difficulty I have described above could be resolved, there would remain one additional, intractable difficulty in the case of gardens: for two identical perceivable

22 However, Currie claims that the aesthetic and art-historical values of Leonardo's and Duchamp's perceivable structures are identical and that the physical works are interchangeable. (See 129–32.)

garden-structures to co-exist, they would need to have been enacted not only in the same location but at precisely the same moment in time and in identical growing condition, which possibility is remote, to say the least.[23] Therefore, while Currie's Action Type Hypothesis claim that artworks have a perceivable structure and a unique heuristic path may be appropriately hypothesized in the case of gardens, its corollary, that another artist could create the "same" garden via a different heuristic path must, in the case of gardens, remain untestable.

Second, Currie's Action Type Hypothesis implies that any instance of the perceivable structure of an artwork is a product of a completed action. It is therefore reasonable to assume that an artist's actions have resulted in her producing a completed work or a completed set of instructions for the performance of a work. This further implies that an artist has foreseen all possible development and changes in non-performance works and has anticipated and found acceptable those developments and changes that might occur in performance works. But this cannot be the case in gardens. Gardens may persist for centuries and are subject to continuous change on all levels, from the geological to the biological. It is not plausible, or possible, to expect all changes in a garden's contents, structure, views, etc. over any other than an extremely short period of time to have been foreseen, approvingly or otherwise, by the garden designer and the Action Type Hypothesis is, on these grounds, inadequate to the case of gardens.

Instance Multiplicity Hypothesis

Currie's Instance Multiplicity Hypothesis claims that the perceivable structure of any artwork is capable of being multiply-instanced.[24] This hypothesis enables him to make the controversial claim that works (perceivable

23 However, it is granted that any number of identical, perceivable garden structures could theoretically exist in theoretical, parallel universes.

24 "Multiply-instanced" is not an endearing term, but I use it to be consistent with Currie and to retain the implications it carries for him. See: Currie, *An Ontology of Art*, 85–129.

structures) that are usually regarded as singular, such as *Mona Lisa* and gardens such as Tupare, are in fact of equal aesthetic value and importance in their original instance *and* in a theoretically limitless number of copies that are physically indistinguishable from the original. In the case of Leonardo's and Duchamp's identical paintings discussed earlier in connection with the Action Type Hypothesis, Currie, in accordance with his Instance Multiplicity Hypothesis, argues that the perceivable structure of Duchamp's work, or any other identical perceivable structure, including, for example, one made by a super-copier, could all stand in for Leonardo's canvas, or vice-versa, without any alteration to whichever perceivable structure's aesthetic or art-historical values.

In Chapter 4 of *An Ontology of Art*, Currie includes a section entitled "The Problem of Architecture," whose contents may fairly be taken, *mutatis mutandis*, to represent his views regarding gardens.[25] In spite of that section's title, architecture turns out not to be so problematic for the author. He admits that replicating architecture, in his instance the Parthenon, is fraught with difficulty, but claims that it is nevertheless possible. That assertion, however, underestimates the art of architecture and, by extension, garden making. For, even when it appears to be otherwise, architectural buildings are not in fact replicable, although some aspects of them of course are.[26] This is so because a building, such as St Mark's Basilica in Venice, is not simply a construction with internal and external faces. Its walls contain and shape internal spaces, but they also contain, shape, and influence *external* spaces, particularly, in the case of St Mark's Basilica, the adjoining Piazza San Marco. In this way, buildings are like sculpture: they animate, form, and influence the spatial environment in which they are placed. This feature is not an "extra." It is an essential function of sculpture, and of architecture too.

All architecture is inevitably part of its geographical setting. Ideally it sets out to merge or contrast with, or otherwise acknowledge its immediate and surrounding geographical setting. All architecture is inevitably also an expression of the cultural, social, and historical milieu in which it

25 Ibid., 104–5.
26 For support of this view see: Davies, *Philosophical Perspectives on Art*, 141–5.

first appears and then persists. Thus, a Venetian palazzo is one thing in the geographical, cultural, social, and historical setting of Las Vegas and quite a different thing in its native Venice, just as a reconstruction of the Sydney Opera House in Moscow would take on resonances different from that possessed by the original on Bennelong Pont in Sydney.

All these relational and situational features of architecture apply to gardens as well. Gardens also contain, shape, and influence the spaces surrounding them, they animate, form, and influence the spatial environment in which they are placed, and they merge or contrast with, or otherwise respond to, their immediate and surrounding geographical, cultural, social, and historical milieux. Because of all these factors, gardens are not capable of being multiply-instanced. However, there are five ways in which gardens are even more formidably unique than architecture.

First, without the interventions of artificial heating, cooling, irrigation, and wind protection, gardens must rely on the resources of the local climate for flourishing. Generally speaking, climates are not replicable with any exactness and for this reason gardens are not replicable on other sites. Although attempts are sometimes made to do this, they often begin with good intentions and end in failure. Second, growth rates and suitability of plants to purpose are governed in part by the geological and pedological make-up of the soil and substrate and the resultant drainage patterns. Such a system is not usually replicable, and for this reason too gardens cannot be multiply-instanced. Third, views out from gardens are often constitutive elements of the gardens that offer them. But such views, of mountains, the sea, rivers, cathedrals, motorways, and so on, are not usually replicable. Fourth, gardens sometimes have indistinct or highly permeable boundaries, or no obvious boundaries at all. In such cases, how is it to be determined how much of the garden's surrounds need to be replicated in order that the garden may be said to have been successfully re-instanced on another site? Fifth, at what stage of development or in what state of health or decline should a garden be duplicated so as to be "faithful" to the original? Is it necessary to plant elm species prone to Dutch elm disease to replicate those dying in the original garden? Should the polluted water in the original garden's lake be duplicated to ensure that fish do not thrive in the replica lake? And so on and on.

The problems raised above do not signal any issues that are unfamiliar to those who have written on gardens and philosophy generally. However, they stand as significant obstacles to Currie's ontological claim that the perceivable structure of any artwork in any art genre is capable of multiple instancing. Currie has not successfully countered the related issues in the case of architecture, nor has he acknowledged or addressed them in the case of gardens. Therefore, his Instance Multiplicity Hypothesis is inadequate to the case of these arts.[27]

Before leaving the issue of the multiple-instancing of gardens, it is relevant to note that attempts are indeed often made, with varying degrees of success, to replicate some features and aspects of gardens and garden styles. Consider, for instance, the "English" gardens of early, white antipodean settlement, and their recently fashionable "Italian" successors. It is also right to acknowledge that some show gardens have been relocated – for example, from the Chelsea Flower Show to the Middle East. However, such cases are exceptional. Show gardens have usually been designed in the first place without primary regard to their geographical, climatic, pedological, and social contexts. Therefore, a relocation of them to another location, in which they may again fail to have regard to those contexts, may not be as problematic as for typical, non-show gardens. Such relocations do not amount to a successful challenge to my earlier claims regarding the multiple-instancing of gardens. Even it is accepted that some gardens *may* be able to be relocated, it still does not entail that they are replicable: in such cases, the relocated garden is not a replica, it is simply the relocated, original garden.

27 There is a further way in which the claim that gardens can be multiply-instanced might be challenged. It might be argued that a particular garden can be multiply-instanced if the original garden is destroyed and a new one, indistinguishable from a particular moment of the original one is installed on the same site. However this claim does not succeed because the replica garden does not represent a multiple instance of the original garden because the original garden no longer exists. The replica garden is thus a (new) singleton, not a co-existing multiple instance of the original. (However, see footnote 23, p. 129, in which I grant that multiple instancing of gardens remains a theoretical possibility in the case of parallel universes.)

Gardens Depend on Physical Objects

Richard Wollheim's *Art and Its Objects* was first published in 1968.[28] In it, Wollheim presented what he called the physical-object hypothesis. In arguing for the hypothesis, he first proposes a common enough division of the arts into one category whose works are singular, such as oil painting and sculpture, and another category whose works typically exist in multiple copies or performances, such as novels and music.

According to Wollheim, singular artworks are identifiable with actual physical objects. Thus, *Mona Lisa* is identifiable with the poplar panel Leonardo painted around 1510 and called "Mona Lisa." This ontological position differed significantly from other contemporary ontologies, especially those referred to by Wollheim as "Ideal," and "Presentational." The Ideal ontology was represented by Croce, Collingwood, and their followers. Their ontology entailed that a work of art was a type of mental production, and that the physical "art" object or event functioned solely as a tool to access an artist's real (mental) work of art.[29] Presentational ontology was represented for Wollheim by the formalist aesthetics tradition, which stretched from Bell to Greenberg and beyond. The Presentational ontology required that anything other than the immediately perceptible aspects of a work be deemed irrelevant to a work's aesthetic interest and value. In the case of extreme formalists, such as Clive Bell, all expressive and representational content was to be ignored. Wollheim, on the other hand, went to some lengths in *Art and Its Objects* to argue how such features of a work could in fact be transmitted via a physical-object artwork and that they therefore did not need to be considered external to the work.

28 Richard Wollheim, *Art and Its Objects: An Introduction to Aesthetics* (New York: Harper and Row, 1968).

29 Had Wollheim been writing later he would surely have included D. Davies' action ontology in his "Ideal" group.

However, in 1980, *Art and Its Objects* was republished.[30] This time it included "Six Supplementary Essays," one of which was entitled "A Note on the Physical Object Hypothesis." In this brief essay, Wollheim declines to support his physical-object hypothesis, writing that "I suspend judgment on its truth." He argues instead for an "'aesthetic object theory,'" claiming that "the likeliest, though not the sole alternative to holding the physical object hypothesis is to posit, for each work of art in question, a further individual, or an 'aesthetic object' with which the work of art is then identified."[31] Thus, a work of art remains for him an object but is no longer identifiable with, say, the physical object – that is, the canvas, applied paint and frame – of the painting Leonardo called "Mona Lisa." *Mona Lisa* has now become an *aesthetic* object.

The concept of the aesthetic object allows for the conceptual safeguarding of what he refers to as the aesthetic character and the aesthetic condition of a work.[32] Acknowledging a work's aesthetic character involves drawing a distinction between what Wollheim now saw as the merely physical, and the aesthetic characteristics of a work of art and acknowledging a work's aesthetic condition involves allowing that a work may be aesthetically privileged at a particular stage of its existence. For example, a bronze may acquire a patina that enhances it aesthetically at the same time as a fresco may become grimy, less aesthetically pleasing and therefore in need of restoration.

Wollheim's acknowledgement that aesthetic objects have constituent, changing character and condition aspects that need accounting for amounts to an (unstated) simultaneous acknowledgement of important ontological aspects of gardens and our experiences of them. Wollheim does not, however, take the opportunity to explore those aspects in the case of gardens (by my reckoning, he devotes 11 words to the subject of gardens). Furthermore, an earlier acknowledgement of these same "new" aspects of artwork-objects would, on its own account, have amounted to an inevitable

30 Richard Wollheim, *Art and Its Objects: An Introduction to Aesthetics*, 2nd edn (Cambridge: Cambridge University Press, 1980).

31 *Art and Its Objects: An Introduction to Aesthetics*, 177.

32 Ibid., 178.

rejection of his original physical object hypothesis if that hypothesis had been tested against the case of gardens. In my discussion below, I examine the notions of character and condition principally in relation to the aesthetic-object hypothesis, on the grounds that Wollheim himself has as good as rejected his physical-object hypothesis.

I begin by quoting Wollheim's 11 words pertaining to gardens. In a section discussing art works which change over time, he cites as an example "or William Kent's garden at Rousham, conceived of with full-grown trees." It is a clause that is of interest because of the three assumptions which underlie it. The first and second assumptions are not controversial: many people, including some philosophers, would agree that gardens can be works of art and that Kent's famous garden at Rousham is such a garden. The third assumption is more controversial, and can be broken down into two smaller elements: gardens are to be viewed as unchanging pictures and not as processes; and, a garden maker has it in mind that her garden will be a work of art only at the time when the trees and other living elements of it are at a particular size and age.

The first element implies that gardens are simply static visual images, or pictures, rather than continuously changing images. However, almost all garden designers and gardeners design and garden in the knowledge that plants and gardens change continuously and that aesthetic interest and value, and other benefits accrue to the gardens of those who are successful in anticipating and managing such changes.

The second element is related to the first. Element one treats gardens as if they are static: element two acknowledges that even though gardens do change there is still a time when they are "at their best," that is, presumably, most like the work of art the designer intended visitors to enjoy. To be absurd about it, it could be argued on this view that Kent's Rousham was/is/will be a work of art only at and for a given moment, and then only if we have evidence of Kent's intentions in this regard. In such an instance, the commonly heard apology from garden owner to visitor that, "You should have been here last week," assumes altogether different proportions.

Before examining Wollheim's aesthetic-object hypothesis in terms of its application to gardens, a brief detour through his physical-object hypotheses is in order because that will provide useful background material when it comes time to discussing his aesthetic-object hypothesis.

Wollheim's physical-object hypothesis is of particular interest to the case of gardens because gardens are generally more "physical" than any other artworks. We can live in a garden and we can eat its produce. Birds, fish, animals, and insects live and feed in gardens. Gardens have many features that are indistinguishable from features found in the non-garden world. Gardens have important elements that are growing, changing, and dying. All these, and many more features of gardens, make them undeniably real in ways and to a degree that most other art objects are not.[33]

But this very realness of gardens, which Wollheim's physical-object hypothesis seemed tailor-made to explain, would have defeated that hypothesis if Wollheim had attempted to apply it. This is so because, with the possible exception of architecture, no other fine art so confuses us with claims of being *simultaneously* part of art and part of life. Gardens objects are in every sense real and "of this world," with everything that entails, at the same time as they are works of art, with all that entails. So, although Wollheim could easily and sensibly have accounted for a garden's being a physical-object artwork, he would have had difficulty attempting to account for the same garden's *simultaneous*, quotidian existence as a physical-object non-artwork.

The case I am making here, that gardens and some architecture are simultaneously "real" objects and works of art, differs from, say, the case of painted canvases being removed from walls and used as stretchers during war time. In the latter case, a painting is mis-used *qua* art by being used as a stretcher. A canvas is expected to be in a frame and on a gallery or church wall. Most art gardens, by contrast, are meant to be used. They are not just to be looked at: they are to be lain in, eaten in, weeded, and so on. It is expected that their shade will be enjoyed and, sometimes, their produce

33 This leads Miller to claim that, "because they live and they incorporate our bodies," gardens' various effects on us may be made "with an unusually compelling force." See Miller, *The Garden as an Art*, 5.

consumed. They are simultaneously real and works of art in a manner unique among art genres.[34]

Wollheim's aesthetic-object hypothesis represents a considerable modification to his earlier physical-object hypothesis. In the case of gardens, his acknowledgement that works of art have both aesthetic and non-aesthetic components, and that these may vary over time, goes some way towards an ontology adequate to their case. However, the hypothesis only goes so far and is, in the end, inadequate to the case of gardens for two reasons.

First, the aesthetic-object hypothesis acknowledges that artworks have aesthetic and non-aesthetic components. Thus, in a garden, we are used to distinguishing between the roses and statuary on one hand and the ice cream van parked just inside the gate on the other. However, the hypothesis is unable to account for the very many, more complex cases where garden elements, and our experiences of them are simultaneously aesthetic and non-aesthetic, or that alternate between being aesthetic and non-aesthetic, or vice-versa, in quick succession or during different seasons. For example, in a garden you can sunbathe, chase a peacock, add your "song" to the bird song, or eat a garden peach and wash your hands in the fountain, all while simultaneously enjoying the garden as a work of art.

Second, and more importantly, the aesthetic-object hypothesis acknowledges that some works of art *do* change and that, in such cases, unless the artist's intentions in this regard are clear, a decision needs to be made about whether such changes are for the better or worse and, therefore, whether works are aesthetically privileged at some or other time of their existence. A stone building may be enhanced by its gradually acquired patina while a timber building's appeal may be lessened by the peeling paint on its woodwork. But in the case of gardens, unless the designer has specifically indicated otherwise, it is fruitless to speculate on whether a garden is better aesthetically at one time than at another because the whole point of gardens is change. Without growth, decay, death, flowering, seeding, furling, unfurling, and so on, gardens cannot exist.

34 I discuss the theories of Danto in this regard at 179–82.

So, Wollheim's aesthetic-object hypothesis turns out to be inadequate to the case of gardens, even though it sets out to address the issues of realness and change that are paradigmatic to the case of gardens. It is inadequate principally because it fails to accommodate the continuous, sometimes unpredictable, but aesthetically important and valuable changeability of gardens' living elements.

Of the three ontologists whose work has been discussed in this chapter only Davies has paid any but the slightest attention to gardens; and his ontology, along with those of D. Davies and Wollheim, remains in part inadequate to their case. It is further noteworthy that "ontology" does not appear in the indexes of Miller's, Ross's, or Cooper's substantial garden monographs. The ontology of gardens remains largely unexplored territory. In Chapter 8, I offer a preliminary account of an ontology of art gardens that celebrates, rather than ignores or renders problematic, their mutability and singularity.

PART IV

Breaking New Ground

CHAPTER 7

A Philosophical Detour: Definitions, Theories, and Ontologies

Definitions and Theories

Philosophers were uninterested in gardens during the nineteenth and early twentieth centuries and during the late twentieth century a small number of them entertained the possibility of rehabilitating gardens as works of art. The philosophers who at those times ignored or sought to include gardens in the category of "art" typically relied on a generic, one-size-fits-all definition of "art." The traditional philosophical preference for such all-encompassing definitions, with their attendant necessary and sufficient conditions, meant that philosophers narrowed the extension of the defined term, thereby excluding objects and activities which might otherwise have counted as "art." Since the time of Plato, who proclaimed all (what we now think of as) art to be a matter of imitation, there has followed a series of equally all-encompassing definitions, that all art is expressive of emotion, or exhibits significant form, or relates to an historical art narrative, and so on. Thus, the notion that a single definition of "art" could somehow accommodate all individual artworks in all genres was taken for granted by philosophers until the mid-twentieth century, and that assumption is still held by some. Furthermore, up until that time, the successive definitions had each, on Davies' terms, fulfilled the dual roles of providing a definition and underpinning a theory of relevance, importance, and value for art.[1]

1 See earlier discussion of this point at 98.

The notion of a single definition/theory of art remained largely unchallenged until the mid-twentieth century, when Morris Weitz published his influential paper "The Role of Theory in Aesthetics" in 1956.[2] Although, as Kivy notes, the American philosopher Dewitt Parker was already questioning whether all art indeed had a common essence as early as 1939, from around the time of Weitz's paper, a growing number of philosophers of art has emerged to challenge the necessity, appropriateness, and usefulness of single one-size-fits-all definitions and/or theories of art.[3] These philosophers may have differing agendas, and arguments, but they all agree that unique definitions and/or theories for individual art genres might usefully be developed.

In "The Role of Theory in Aesthetics," Weitz challenges the traditional historical definitions of "art." He rejects mimesis, formalism, and expressionism, among others, on the grounds that while each has purported to be a complete definition of "art," each has necessarily rejected artworks, genres, and styles that were non-compliant with the definition. Invoking Weitz's arguments, one might say that although Palestrina's masses are universally acknowledged to be works of art, according to the contemporary mimetic definition they cannot be works of art because they are generally acknowledged to be non-mimetic.

Weitz argued that each of the historical definitions was partially correct and partially incorrect and that each necessarily rejected conflicting evidence to give an illusion of completeness. Each definition assumed a "closedness" of the concept, which he argued was inappropriate for the concept of art. He rejected the possibility of *defining* anything other than the constructed and closed concepts of logic and mathematics and argued that art's creative dimension made it importantly open to change and novelty, and that it was therefore particularly undefinable.

Instead, Weitz claimed that art had an "open texture," which it shared with the term "game," and this similarity with "game" enabled him

2 Morris Weitz, "The Role of Theory in Aesthetics," *The Journal of Aesthetics and Art Criticism* 15, no. 1 (1956).

3 Peter Kivy, *Philosophies of Art: An Essay in Differences* (Cambridge: Cambridge University Press, 1997), 32.

to introduce the Wittgensteinian notion of family resemblance, which Wittgenstein originally developed using "game" as his exemplar.[4] Weitz proposed replacing all definitions of art with a description of the family resemblance between the *arts*. Such a description would reflect the non-essential nature of the arts, acknowledge their "complicated network of similarities overlapping and crisscrossing," and reflect their possession of "no common properties – only strands of similarities."[5]

Weitz's rejection of the historical definitions of art did not involve a simultaneous rejection of art theory. His rejection of the historical definitions was purely on the grounds of their being inadequate as definitions and he accepted their putative usefulness as theory. He wrote: "To understand the role of aesthetic theory is not to conceive it as definition, logically doomed to failure, but to read it as summaries of seriously made recommendations to attend in certain ways to certain features of art. ... [Aesthetic theory] is of the greatest importance to our understanding of the arts."[6]

In these circumstances, it is ironic to consider that during the 50 or so years following Weitz's paper, new definitions of art's essence have appeared at an historically unprecedented rate. Several of these definitions, including those put forward by Danto, Dickie, Levinson, and Carroll, embraced the challenge of incorporating the radical creativeness that had challenged Weitz. In fact, that radical creativeness was the generating impulse for at least some of these philosophers' definitions of art's essence.

The first of the post-Weitzian definitions did not aim to be inclusive of the products of the radical avant-garde. It appeared in 1958, in Beardsley's *Aesthetics: Problems in the Philosophy of Criticism*, and defined art in terms of aesthetic functionalism.[7] Beardsley claimed that something was a work of art by virtue of its ability to provide an aesthetic experience: a work of art was "an arrangement of conditions intended to be capable of affording an

4 Weitz, "The Role of Theory in Aesthetics," 31.

5 Quotations from Wittgenstein's *Philosophical Investigations* appear in: ibid., 30–31.

6 Ibid.

7 Monroe C. Beardsley, *Aesthetics: Problems in the Philosophy of Criticism*, 2nd edn (Indianapolis, IN: Hackett Publishing Company, 1981).

experience with marked aesthetic character."[8] This was followed by Danto's definition in his 1981 volume, *The Transfiguration of the Commonplace.*[9] It sought to establish why a seminal "work of art," such as Warhol's *Brillo Boxes (1964)*, was art whereas its quotidian equivalent – mere Brillo boxes – were not.[10] Danto claimed that art was to be distinguished from (perceptually indistinguishable) non-art by virtue of its "aboutness." "The former," he wrote, "are about something (or the question of what they are about may legitimately arise)."[11] Next came Dickie's institutional definition. Although later modified from its initial publication in 1974, its central premise, that art was to be defined not by any intrinsic qualities it might have but in virtue of the relationship the would-be art object or event enjoyed with the art-world, remained unchanged.[12] The importance of a work's relatedness was retained in definitions offered subsequently by Noël Carroll, and Jerrold Levinson. In Carroll's case, a work of art was an object or event that stood in a particular relationship to its artistic predecessors and, in Levinson's case, it was an object or event that could be fitted into a coherent narrative with past art. These relational definitions have been followed more recently by the disjunctive definition proposed by the philosopher Robert Stecker, in which works of art may either exhibit aesthetic functionalism, *or* they may satisfy the requirements of Dickie's Institutional definition, *or* they may exhibit the relatedness to the past that Carroll or Levinson requires. And finally came the cluster concept definition, promoted most notably by Berys Gaut. Gaut proposed a list of 10 criteria, none of which is necessary but some (one?) of which is sufficient for something's being a

8 Ibid., 299.

9 Danto, *The Transfiguration of the Commonplace: A Philosophy of Art.*

10 Brillo is a well-known brand of household cleaning products in the United States. "Brillo boxes" refers to the boxes in which the Brillo products are packaged, transported, and displayed. The title, *Brillo Boxes (1964)*, refers to artist Andy Warhol's work, which comprised an arrangement of hand-made Brillo boxes that looked in all respects identical to their machine-made, everyday equivalents. See: Andy Warhol, *Brillo Boxes, 1964*. Stable Gallery.

11 Danto, *The Transfiguration of the Commonplace: A Philosophy of Art*, 82.

12 George Dickie, *Art and the Aesthetic* (Ithaca, NY: Cornell University Press, 1974), 31.

work of art.[13] His list includes, for example: "possessing positive aesthetic qualities"; "belonging to an established artistic form"; and, "being formally complex and coherent."[14]

I have listed all these definitions because, with the exception of the cluster concept, whose usefulness has otherwise been questioned on the grounds that a theoretically limitless number of criteria might be added to Gaut's list without destroying its intent, the definitions have all received convincing counter-arguments concerning what each of them has defined as the essence of art. Therefore, it appears that despite the flurry of definitional activity over the past 50 years, the situation is little changed from that described by Weitz in 1956: any definition purporting to be a complete definition of "art" must necessarily end up rejecting artworks, genres, and styles that are non-compliant with that definition and therefore be inadequate as a definition of "art."

An alternative approach to answering the question of what a work of art is involves enquiry into art's, and the arts', *nominal* essence(s). Three philosophers working in this field, are Peter Kivy, Dominic McIver Lopes, and Aaron Meskin. A plausible extrapolation from their writings is that the way forward in the areas of the definition, theory, and ontology of art lies not in developing or refining a single, overarching, multi-art-genre definition, theory, or ontology but in devising singular, genre-specific definitions, theories, and ontologies.

Kivy uses the case of abstract music to construct a fascinating historical narrative of the history of definitions of art from the eighteenth century onwards.[15] Invoking Carroll, who writes that "one might say that a great deal of modern philosophy of art is an attempt to come to a philosophical understanding of the productions of the avant-garde," Kivy explains how the rise of the art theories of expressionism in the early nineteenth century and of formalism in the early twentieth century were in their own ways as much attempts to grapple with contemporary avant-gardes as were the

13 B. Gaut, "'Art' as a Cluster Concept," in *Theories of Art Today*, ed. N. Carroll (Madison: University of Wisconsin Press, 2000).

14 Ibid., 28.

15 Kivy, *Philosophies of Art: An Essay in Differences*.

definitions of Dickie, Carroll, and other philosophers during the later years of the twentieth century.[16]

Kivy cites the failure of formalism *à la* Clive Bell and Roger Fry to accommodate adequately what he calls contentful (i.e. representational in the conventional sense) art as a turning point in the quest for a single definition of the essence of art. "It seems to me that at this point in the history of aesthetics one might well have begun to doubt seriously whether a single theory, a single real definition, could lasso both the contentful arts and absolute music together in the same modern system."[17] According to Kivy, the search for the Abbé Batteux's *même principe* for all the arts foundered at that time and needed to find new directions.[18] One of these directions, the Weitzian one, I have already described, and it involved moving away from the notion of defining "art" altogether. The other direction, pursued initially by Danto and Dickie, involved a continued search for *un même principe* for the arts, but their search, and that of some of those who followed, was for a completely different sort of property – an extrinsic, relational property – that might serve to unify the arts.

Kivy's response to this dilemma has been to develop philosophies of *individual* arts, most notably literature and music.[19] He concludes that the search for a unifying principal, while useful on its own account, "has blinded the philosophical community to a bevy of questions of more than trivial importance, involving the arts not in their sameness but in their particularity."[20]

Lopes argues along different lines but reaches a similar conclusion in "Nobody Needs a Theory [definition] of Art."[21] He claims that the two

16 Noël Carroll, "Historical Narrations and the Philosophy of Art," *Journal of Aesthetics & Art Criticism* 51 (1993), 314. Quoted in: Kivy, *Philosophies of Art: An Essay in Differences*, 6.

17 *Philosophies of Art: An Essay in Differences*, 28.

18 For a brief discussion of Batteux's treatise, see 50–1.

19 For example, see: Peter Kivy, *The Performance of Reading: An Essay in the Philosophy of Literature*, New Directions in Aesthetics (Oxford: Blackwell, 2006); *Introduction to a Philosophy of Music* (Oxford: Clarendon Press, 2002).

20 *Philosophies of Art: An Essay in Differences*, 53.

21 Lopes, "Nobody Needs a Theory of Art."

main streams of definitions of "art" in the twentieth century, that is the functionalist and the proceduralist streams, each evolved in response to the provocations of avant-garde works. To the aesthetic functionalists, these provocative works were ruled to be beyond the art pale. For the proceduralists, the same works provided the stimulus to revolutionize the concept of art by claiming that certain of art's non-perceptual qualities were to be the basis on which something was deemed to be or not be art, and therefore, according to the revised "art" concept, those same provocative works were deemed to be works of art. Lopes argues that while each of these definitional streams has provided new and useful insights, the better way to have dealt with the troublesome avant-garde works would have been through the development and application of theories of the individual arts and individual artforms. For instance, Duchamp's *Fountain* might more usefully have been considered as a visual or sculptural artwork, and Cage's *4'33"* as a musical work, rather than their being considered as art *simpliciter*. He claims that "there is nothing left to explain when you have [adequate] theories of the art forms and the arts," and that the subject needs to be changed from art to arts and artforms.

Further argument against the art definitional project is provided by Meskin.[22] He agrees with Weitz concerning the possibility and utility of defining "art." He suggests that, as the "most plausible" of the contemporary (i.e. post-Weitzian) definitions of art have given no guidance on the important matters "of interpretation, evaluation and appreciation" of artworks, it may be reasonable to propose that definitions of the individual art forms might be more useful in these respects.[23] He points out that Weitz himself suggested that "many of art's various sub-concepts (i.e., the art form and genre concepts) cannot or should not be defined"[24] and with this suggestion Meskin also agrees. Meskin proposes that the various art forms may

22 Aaron Meskin, "From Defining Art to Defining the Individual Arts: The Role of Theory in the Philosophies of Arts," in *New Waves in Aesthetics*, eds Kathleen Stock and Katherine Thomson-Jones (Houndmills, Basingstoke, Hampshire; New York: Palgrave Macmillan, 2007).

23 Ibid., 143.

24 Ibid.

indeed be indefinable – or at least not usefully definable in terms of their unique values, processes, meanings, and so on.

Meskin's claim is that neither "art" nor "arts" can be usefully or profitably defined. He cites Zangwill: "This issue does not concern the word 'art'. … We want to know about a range of objects and events, not about the words or concepts we use to talk about those things. We are interested in objects, not concepts – the world not words."[25] Meskin asserts that definitions of art and of the various arts are not, and never have been, necessary for adequately and legitimately experiencing, evaluating, and criticizing works of art. Echoing Kendall Walton (see below), he claims it is sufficient that we know "enough about how the various art forms, genres, and styles work." And finally, he concludes: "We may need theories of the arts but I do not see any reason to think that we need definitions."[26]

If, then, Kivy's, Lopes', and Meskin's shared claims, that theories, not definitions, should be developed for individual artforms and genres, what should the characteristics of such theories be? How are we to be guided in developing a theory of, for instance, gardens? What sort of theory would adequately account for Tupare garden, presented in Chapter 1? I believe that an important and widely read paper of the American philosopher Kendall Walton offers some direction in answering these questions.

While Weitz's "The Role of Theory in Aesthetics" may be claimed as a direct, if distant, precursor to the work of Kivy, Lopes, and Meskin with respect to the arts' nominal essence(s), Walton's 1970 paper, "Categories of Art," may be seen as an indirect, though important, contributor to the sympathetic philosophical environment within which those philosophers could develop their claims. Although the aim of Walton's paper was to provide support for a limited intentionalist position, the paper's examination of the unique combinations of properties characteristically present and absent in works in a given art genre, and the psychological claims flowing from that examination, can be seen as adding weight to Weitz's quest for new ways of theorizing about art.[27] Walton's paper can be taken to endorse the view that theories of art that do not account for the unique and uniquely

25 Quoted in: ibid., 139.

26 Ibid., 143–4.

27 Kendall Walton, "Categories of Art," *The Philosophical Review* 79, no. 3 (1970).

important features that are characteristic of different art forms and genres are not useful. His paper also indirectly questions what characteristics an adequate account of a specific artform might possess and for which characteristics of that art form an account might need to have regard.

Walton offers an alternative to the projects of defining or developing classificatory theories of art, or individual arts. Instead of definition(s), his project concerns normative and evaluative aspects of the arts. He is concerned with determining and categorizing the aesthetically relevant properties of the different arts and thereby providing a psychological framework for the *appreciation* of artworks. His paper indirectly facilitated the replacement of the quest for establishing the unifying factor(s) of all the arts with an acknowledgement, even a celebration, of the potential differences of method, materials, processes, and so on, that different art genres, styles and even, in some cases, individual works, exhibit or possess.

Walton's aim in "Categories of Art" was to counter the claims of Wimsatt and Beardsley's "The Intentional Fallacy" by arguing that a work's aesthetic properties are not solely discoverable in the physical work itself.[28] He wrote: "If a work's aesthetic properties are those that are to be found in it when it is perceived correctly, and the correct way to perceive it is determined partly by historical facts about the artist's intention and/or his society, no examination of the work itself, however thorough, will by itself reveal those properties."[29]

In arguing for this claim, Walton elaborated in detail on what he described as "standard," "variable," and "contra-standard" properties of works of art in a given genre. I will not expand on these properties here other than to say that in recognizing them in a work of art that we are experiencing we are, albeit almost always unconsciously, recognizing that we are in fact hearing a sonata; and, we do not mistakenly think that we are reading a poem, for example, or looking at a painting, or watching a ballet. Walton argued that this was the case because certain properties of, say, a poem, are standard, variable, and contra-standard for poetry and that these properties, although not perceptually present in the poem, need to

28 W. K. Wimsatt Jr and M. C. Beardsley, "The Intentional Fallacy," *The Sewanee Review* 54, no. 3 (1946).

29 Walton, "Categories of Art," 363–4.

be taken account of if we are to appropriately experience or criticize that poem. It is not appropriate, for instance, to experience that poem as a sculpture because sculptures possess different, though still non-perceptual in the work, standard, variable, and contra-standard properties than poems do. In summary, he claimed that we will be categorizing a work appropriately, that is, as a poem rather than a sculpture, if (a) it has a relatively large number of properties standard for the category of poems, (b) the work "comes off best" when experienced as a poem, (c) the artist intended it to be perceived as a poem, and (d) poetry was a recognized art genre in the society in which it was produced.

A straightforward example will clarify Walton's claims. We are right in experiencing Anthony Trollope's *Barchester Towers* as a novel. It has many features standard for a novel, such as being a prose narrative, having chapters, having a plot and various characters, etc. It has some features variable for a novel, such as its comic elements, its fictitious locations, and its exposition of ecclesiastical intrigues. I am not aware of any contra-standard properties it possesses, although the presence of pages of musical notation, or the author's requirement to stroke or smell the pages while reading would qualify as such. Furthermore, *Barchester Towers* works well as a novel, Trollope intended it to be a novel, and novels comprised a well-known art genre in the Victorian period. So, for all these reasons, our experience and criticism of *Barchester Towers* as a novel is appropriate even though the evidence supporting these reasons is not able to be perceived *in* the novel. In this way, Walton was able to counter the anti-intentionalist claims of Wimsatt and Beardsley.

Walton does not discuss gardens. Had he done so, it is reasonable to assume that he would have sought to establish their standard, variable and contra-standard properties, and the role those properties play in appropriate experiences and theories of gardens. In other words, if Tupare is to be categorized as a Waltonian "garden," what does that tell us about Tupare and its aesthetically relevant features, and how we experience and evaluate them.[30] In Chapter 8, I claim that the aesthetically relevant features of

30 I discuss Tupare in some detail in Chapter 1.

Tupare, and all gardens, are intimately bound up with their ontology, as, therefore, are the reasons why we value them and find them interesting.

Ontologies

Commentary on the ontology of art and art forms has been occurring since the time of the Ancient Greeks but it was not until the twentieth century that it came to assume a prominent role in philosophical aesthetics. Just as in the case of art's definition, this flourishing of interest in art's ontology was stimulated in part by troublesome products of the twentieth-century avant-garde and today, in part because of the continuing challenges of the avant-garde, the ontology of art represents one of philosophical aesthetics' most important and contested areas of endeavour.

Unlike definitions and theories of art, which seek to establish under which conditions works of art exist, the ontology of art seeks to establish what sort of *entity* is a particular thing, or type of thing, which has already been determined to be art.[31] It considers what material(s) and form(s) constitute the work of art, and the mode(s) in which the work exists. It determines "the conditions under which a work of art comes into existence, remains in existence and is destroyed and also the conditions under which works of art are *one and the same*."[32] It considers whether works of art are physical, imaginary, or abstract objects, or whether they are actions, and whether the identities of works of art are physical, mental, or imaginary.

The ontology of art is of important purely philosophical interest and value, especially in its relationship with metaphysics. However, it can also be practically important and useful and, in this vein, Davies writes that "some classifications and interests are likely to be more revealing of why

31 Amie L. Thomasson, "The Ontology of Art," in *Blackwell Guide to Aesthetics*, ed. Peter Kivy (Hoboken, NJ: Wiley-Blackwell, 2009), 78.

32 Amie L. Thomasson, "Debates About the Ontology of Art: What Are We Doing Here?," *Philosophy Compass* 1, no. 3 (2006), 245.

and how art is created and appreciated ... [and that] ... it is these that our ontology should reflect."[33] Such practical applications may include: guiding viewers, conservators, and valuers towards knowing how much a painting can change through deterioration or restoration and still remain the same work; in the case of less than word- or note-perfect performances of plays and musical works, guiding critics and recording producers towards knowing, say, how many wrong notes are needed before it can be said that a particular symphony is no longer being performed; establishing which properties of a work are essential and which are accidental; establishing how a work is to be displayed; and establishing what aspects of a work it is appropriate to evaluate and interpret.[34]

The recent history of the ontology of art exhibits three trends which I call "conservative," "radical," and "innovative." The conservative approach, exemplified by Wollheim, and Wolterstorff, among others, acknowledges that some works in some art genres, such as painting and non-cast sculpture, are to be identified with their physical objects, and that works in some other genres, such as music and literature, are abstract types, of which individual performances and instances are tokens. I call this approach conservative because it rightly acknowledges that one single ontological theory is inadequate to capture the highly diverse range of artworks and genres.

The radical approach, exemplified by, for example, Currie and D. Davies, claims that one single type of ontological entity encompasses all artworks and genres. In Currie's case, his theory requires that all artworks are *enacted*. They are constituted not by any particular tangible object but by the combined experience of the physical object and the artist's heuristic path in creating it.[35] I call this approach radical because it ignores any differences between art genres and contradicts the widely held, common sense notions of what almost all artists, critics, and the public hold to be intuitively "true" about art objects and their experiences of them. It has been said that we should reject such radical ontologies, "which would

33 Davies, "Ontology of Art," § 1.

34 Thomasson, "Ontological Innovation in Art," 119.

35 See my discussion of D. Davies' ontology in Chapter 6.

take everyone to be massively mistaken about … [artworks'] identity and persistence conditions."[36]

The innovative approach is most readily associated with the work of Thomasson. I will not examine her claims for works of art being abstract or concrete artefacts.[37] My interest is primarily in the grounds on which she has judged existing ontologies of art to be inadequate, the qualities which she claims successful ontologies need to have, and the issues successful ontologies need to address.

Thomasson is clear that "'What is the ontological status of the work of art?' is an ill-formed and unanswerable question."[38] She claims this to be the case on the grounds that "art" seems not to be "category specifying."[39] She further claims that ontology has lost its way and that a preference for what she terms the "discovery" mode of acquiring ontological knowledge has lead us "badly astray."[40]

Her solution to these apparent dilemmas is to propose that we have got things the wrong way about. Instead of beginning ontological investigations by bringing traditional ontological conceptions to bear on art we should, for a start, make our investigations genre- if not work-specific, and then investigate that genre's or work's nature while not treating it as some mind-independent thing. "Instead, we must follow the method of analysing the conception embodied in the practices of those competent speakers who ground and reground reference of the term."[41] In this regard, "would-be grounders have some forms of epistemic privilege," and this generally depends on "background practices already in place that co-evolve with the use of the art-kind term."[42] It is only at that point that relevant "features may then be drawn out in more formal philosophical theories of the ontology

36 Thomasson, "Debates About the Ontology of Art: What Are We Doing Here?," 250.

37 Ibid., 247.

38 Ibid., 250.

39 Ibid.

40 Amie L. Thomasson, "The Ontology of Art and Knowledge in Aesthetics," *The Journal of Aesthetics and Art Ciriticsm* 63, no. 3 (2005), 221.

41 Ibid., 226.

42 Ibid., 225.

of the work of art – explicitly describing … their relevant formal category, … existence conditions and relations to human intentions and physical objects and processes, and their boundaries and individuation conditions."[43]

Thomasson is arguing for an ontology of art which, at least in its initial formulation, results from a bottom up rather than a top down process. She argues that more accurate and appropriate ontologies of the arts will emerge from studies of the practices and unspoken precepts, traditions, and so on that are reflected in the art activities, and their practitioners and critics, than will emerge from massaging and manipulating art works and genres into categories that ill fit them and that inadequately explain, or contradict, the ontological qualities that "competent speakers" assume them to possess.[44]

Thomasson concludes that we need to "develop broader and finer grained systems of ontological categories," not only to better understand the traditional arts but also to come to terms with the ontologies of new artforms such as internet and street based arts."[45] I would add to this the need to develop adequate ontologies for art genres that may not be new but which have been previously largely ignored in this respect including, especially, gardens.

Definitions, Theories, Ontologies, and Gardens

For the two contemporary philosophers most prominently involved in writing about art gardens, the unique qualities such gardens possess are problematic. Both Miller and Ross wrote their seminal works at a time when there still existed in some parts an assumed "requirement" to consider and assess any art form in terms of its fit with a pan-art conception

43 Ibid., 225–6.

44 Thomasson is not alone in thinking along these lines. For example, Davies writes of the novel: "Just what is crucial to a story's or novel's identity is likely to be settled by reference to its genre and the literary practices, conventions, and histories on which it draws." (Davies, "Ontology of Art," § 2.3.)

45 Thomasson, "The Ontology of Art," 88.

of art.[46] Without the tasks of (Miller) proving that in spite of definitional problems gardens can be art and (Ross) showing that gardens were art in the eighteenth century but not since, their important books would be considerably smaller. In other words, they both spent time massaging gardens into or out of the (an) "art" conception. By not worrying about whether or not gardens measured up to "art," and in the light of more recent theorizing, they may well have structured their important books differently. To a degree, by simply replacing "problem" in their texts with "attribute," this chapter becomes in part redundant. However, the task *is* more complex than that, and important "new" dimensions of gardens become evident when the model of "art" is abandoned as the ultimate criterion of conceptualization and assessment of garden art.

In her Introduction, Ross writes that her book addresses the "fundamental question of aesthetics, namely, 'What is art [singular] and what does it do for us?'," and her book is strongly focused on assessing how gardens do or do not measure up to "Art."[47] Later in the book, although she argues that the influence of *ut poesis pictura* "waned after the publication of Lessing's Laokoön (1766), which argues that painting and poetry serve separate tasks," she nevertheless persists with what is for her a fundamental question: "In asking what gardens can do and what they can be, I am really asking in what ways they resemble their sister arts."[48] She describes the by-now-familiar range of gardens' "problems" for the "art" conception, thereby perhaps acknowledging inexplicitly the inadequacy of that conception to the case of gardens.[49] In similar vein, Miller titles the relevant sections of her book "Problem 1," "Problem 2," and so on.[50]

Writing later, Salwa is explicit in his dissatisfaction with the use of non-garden models for conceptualizing gardens. With reference to poetic

46 See: Miller, *The Garden as an Art*; Ross, *What Gardens Mean*.

47 *What Gardens Mean*, xii.

48 Ibid., 50, 24.

49 The by-now-familiar range of gardens' "problems" includes: the lack of authorial control; the mutability of the garden object; the environmental, immersive, multi-sensual character of gardens; the naturalness of gardens' materials; gardens' spatial/temporal nature; gardens' combination of art and nature; gardens' functionality and potential usefulness; the non-art interest in and value of gardens' materials; and so on.

50 Miller, *The Garden as an Art*, 73–91.

and painterly models, he writes that "they result in a reductive enclosing of gardens in a grid of concepts offered by art-centered aesthetics that remain blind to those aspects that can be grasped by aesthetics of nature," and he goes on to develop his performance-based model for conceptualizing gardens.[51]

The architect and garden writer Jan Birksted discusses the requirements for an adequate account of gardens and landscapes.[52] He contrasts the theoretical and disciplinary requirements of a garden history-and-theory discipline with the requirements of those he terms adjacent disciplines, including, in particular, art history. His requirements confirm and extend those expressed implicitly and explicitly by the philosophers discussed above, whose concern was for the discipline of philosophy alone. Concerning the documentation and re-presentation of gardens, with which philosophy and garden theory both necessarily engage, he writes: "The risk is converting landscape and gardens into documents without dealing with their particularity and specificity as presentations – which are not purely visual nor simply two-dimensional and static."[53] He claims that the "adjacent disciplines have their own particular methodological approaches, theoretical positions and explanatory rationales, which tend not to take into account the specifics of landscape [or the garden] – its diverse materialities, complex visualitites, [and] composite dimensionalities."[54] Therefore, "the challenge that falls to landscape and garden history and theory is to devise ways of dealing with observation and analysis of domain-specific form and representation when considering the (extra)visual and mobile beholder in landscapes and gardens' three-dimensional space and time."[55]

Landscape architect Bernard Saint-Denis and garden historian John Dixon Hunt offer perspectives on the matter from their respective disciplines. Saint-Denis seeks the essential nature of (all) gardens, and thus their distinctiveness from other arts, by contrasting the classically inspired Italian

51 Salwa, "The Garden as a Performance," 385.
52 Birksted, "Landscape History and Theory: From Subject Matter to Analytic Tool."
53 Ibid., 8.
54 Ibid., 6.
55 Ibid., 8.

garden of the Villa Lante with Gilles Clément's naturalistic *jardin en mouvement* at Parc André Citroën in Paris.[56] Hunt suggests that his own tasks as an historian are made more difficult by the tendency of philosophers to "parry and thrust over definitions of garden art without much commitment to its actual messy, material, and changeful world."[57]

Implied in the claims of Birksted, Hunt, Saint-Denis and, to a lesser degree, those of Ross, are ekphrastic concerns. Words, static images, and film cannot adequately re-present or describe a garden or garden experience, and the interpretation of gardens by way of some other art form (e.g. poetry, painting, or performance arts) may be illuminating in the cases of some feature or style of gardens but remains an inadequate re-presentation or description of an actual garden. Furthermore, considering the process in reverse, attempts to design gardens based on paintings or music may produce effective gardens but, inevitably, the essence of the paintings or music gets left behind in the process. Julie Moir Messervy's *Music Garden* in Toronto takes as its inspiration Bach's *Suite No. 1 in G Major* for unaccompanied cello *and* a performance of the suite by cellist Yo Yo Ma,[58] but the garden is necessarily unable to reproduce the sonic experience which is the *sine qua non* of any performance of the Bach *Suite*. Similarly, the *illusion* of three-dimensional space which, according to Langer, characterizes the painting art form, is necessarily lacking when Jarman returns to the

56 See: Bernard St-Denis, "Just What Is a Garden?," *Studies in the History of Gardens & Designed Landscapes* 27, no. 1 (2007).

57 Hunt, *Greater Perfections: The Practice of Garden Theory*, 7. However, philosophers may choose to take some comfort from art historian James Elkins, who proposes that garden writing may be vague and reverie-like because gardens themselves may possess a "dreamy quality" and thus inspire a sympathetic style of writing and "analysis." See: J. Elkins, "On the Conceptual Analysis of Gardens," *Journal of Garden History* 13, no. 4 (1993).

58 For a description of the garden see: "A Lyrical Landscape: Toronto's Waterfront Music Garden Bridges All the Senses," *Country Living Gardener*, January-February 2003, 72.

themes of his *Avebury Paintings* in the *actual* three-dimensional designed spaces of his well-known garden at Dungeness.[59]

In sum, it is necessary to develop an artform-specific account of gardens. Existing accounts undersell gardens when they measure them against the "art" criterion. Such accounts necessarily ignore, misrepresent, or treat as problematic many of gardens' unique and essential qualities, and the ways in which we experience them. With the exception of Cooper, it is made clear by philosophers writing on gardens that it is only with some difficulty that gardens fit into pan-art accounts of art.[60] Miller admits there is a problem when, commenting on the perceived non-acceptance of some paradigmatic "art" gardens into the fold of fine arts, she writes that "we're obviously doing something wrong here."[61] Salwa claims that gardens are not "standard artworks" for philosophers because they contain "too much nature."[62] And he adds, more forcefully:

> if we – as it is traditionally done – take painting, poetry or architecture as our points of reference, then we can, indeed, state in the Platonic vein that gardens lack an artistic essence for not only do they have no structure typical for arts, but also they are too ephemeral, changeable or unstable to by analysed in any way and thus to have any "conceptual foundations."[63]

Gardens comprise a highly various art form. The conceptual and physical distances between, say, Schwartz's *Splice Garden*, which contains a small

59 See: Michael Charlesworth, "Derek Jarman's Garden at Prospect Cottage, Dungeness, and His Avebury Paintings," *Studies in the History of Gardens & Designed Landscapes: An International Quarterly* 35, no. 2 (2015).

60 Cooper swims against the time when he claims that the art status of gardens is philosophically uninteresting because the questions it raises are "too close to similar and familiar ones asked about other artworks to raise [any] novel issues." (See his *A Philosophy of Gardens*, 12.)

61 Miller, *The Garden as an Art*, 91.

62 Salwa, "The Garden as a Performance," 375. However it is necessary to begin by rejecting the idea, implicit in the quotation from Salwa, that there is any such thing as a standard artwork. It is hoped that one outcome of this chapter will be to show that gardens, at least, comprise a unique or, if you prefer, non-standard art form.

63 Ibid., 377.

number of highly stylized plants, all of which are plastic, and the high-summer lushness of Monet's garden at Giverny can scarcely be ignored. Although those gardens do share qualities – for example, both are designed external spaces characterized by the use of natural materials, an openness to the sky, and Miller's "excess of form" – they exemplify quite different aesthetic issues.[64] In *Splice Garden*, one might be prompted to question whether a work of garden art can be (virtually) static and contain only plastic plants and, at Giverny, one might question how much seasonal change it is possible for a work of (garden) art to sustain and still remain conceptually the same artwork.

Gardens' materials exhibit a degree of ontological hybridity. With reference to the hybrid art of ballet, Davies writes that it "is not unusual … for displaying a range of not clearly differentiated ontological types," and that "even among non-hybrid artforms, most are ontologically various."[65] Within a single garden, individual components of differing ontological types may co-exist. In a typical Italian Renaissance garden, some constructed components of the garden function ontologically in the manner of relatively static architectural or sculptural features and some natural features, such as topiary trees and shrubs, function similarly; whereas fruiting trees, the changing planting infill in parterres, and animated water are relatively ephemeral elements whose ontologies are more easily conceptualized in non-architectural, non-static ontological terms, such as those associated with dance, music, or mobile sculpture.

Gardens are also highly various epistemologically. They can be understood and experienced, with varying degrees of appropriateness and with the caveats discussed in earlier chapters, as paintings, poems, architecture, sculpture, and performances, or even, simultaneously, or sequentially, in combinations of these ways. In gardens such as Stourhead, a visitor can appropriately attend to statuary, architecture, and set-piece and changing painterly scenes, to written, poetic, and emblematic inscriptions, and to

64 See my discussion of Miller's definition of gardens at 9–10.

65 Davies, "Ontology of Art," § 6.

multi-sensory stimuli including the kinaesthetic experience of progress around the garden's circuit, with its contested metaphorical associations.[66]

The kinaesthetic modes of experience, whether real or imagined, and appreciation have not gone unremarked in the case of Stourhead, but they are of fundamental interest in the case of all gardens that are able to be physically entered. Along with the olfactory, tactile, and auditory senses, the importance of kinaesthesia as a component of garden understanding and experience has typically been underemphasized, if not ignored. One reason for this is the ekphrastic challenges the experiences of those senses pose for written or verbal description and analysis.[67] Such challenges are certainly greater than those described above with reference to the painterly, architectural, and poetic experiences of garden.

66 See: John Dixon Hunt, "Emblem and Expression in the Eighteenth-Century Landscape Garden," *Eighteenth-Century Studies* 4, no. 3 (1971), 295–310; *The Figure in the Landscape*; "Stourhead Revisited & the Pursuit of Meaning in Gardens," 328–38.

67 See, for example, the acknowledged inadequacies of my description of Tupare at 20–2.

CHAPTER 8

A New View of the Garden

Introduction

In Chapter 1, I analysed the garden of Tupare with respect to the issues it raises for philosophical aesthetics. Those issues can be divided into two categories: definitional and non-definitional. The definitional issues related to Tupare's arthood and the artist status of its maker. The non-definitional issues related to the garden's ontology and temporality and our experience of and epistemic access to the garden. In this chapter, I present a new, partial theory of gardens that addresses the principal non-definitional issues raised with respect to Tupare, a theory that, in other words, explains how and why Tupare and other gardens function and why they are of interest and value to us. The account focuses on the ontology of gardens' and their materials, and the experiential and epistemological consequences that flow from that ontology.

Ontology

Davies separates all art into two primary ontological categories: works are either singular or multiple in their instances. He subdivides the multiple works into two partially overlapping sub-categories: multiple works are either for performance or not; and, multiple works are transmitted either via exemplars or via sets of instructions. "These ontological divisions," he writes, "are fundamental to the ways we conceive and describe art."[1] How might gardens be accommodated within Davies' ontological categories?

Gardens Are Not Multiple

Gardens have similarities with some multiple artforms, including multiply-instanced works, such as woodblock prints and cast bronzes, and works for performance. These similarities have been commented on by Davies, and Salwa, both of whom imply that gardens may, with some qualifications, be multiply-instanced. A reasonably inferred claim from Davies is that gardens may be multiply-instanced in the manner of, say, woodblock prints or cast sculpture, and Salwa claims that gardens are akin to performances.[2] I reject their claims.

In *Philosophical Perspectives on Art*, Davies considers the possibility of architecture (buildings) being artworks with potential for multiple instancing.[3] He writes that, in a very small number of cases, builders and other tradesmen could construct more than one building from the same set of drawings and that the "copied" buildings and their settings could be virtually indistinguishable from each other, and from an original building if there was one. He suggests that, in that limited sense, buildings could be performances enacted by performer-builders, performer-plumbers, and so on, in accordance with the architect's drawings. However, he rejects this idea on the grounds that neither builders nor plumbers understand their relationship to the drawings and the product to be that of a creative interpreter in the way that, say, an actor playing the role of Hamlet understands his relationship to Shakespeare's text and the play's performance. Davies therefore decides that when buildings *are* capable of being multiply-instanced they are more akin to works in non-performance art genres, such

1 Davies, "Ontology of Art," § 1.

2 Davies makes his claims in terms of architecture but it is reasonable to consider them in the context of gardens, as I do in the following paragraphs.

3 Davies, *Philosophical Perspectives on Art*, 139.

as lithographs and moulded bronze sculptures, the production of which requires executants who are skilled but are not typically regarded as "artists" of the objects they produce. I would add to Davies' reasons for rejecting the notion of performance in architecture the fact that it is the finished, unchanging building that is generally considered to be the putative artwork, rather than the process (performance) of its erection.

I believe there are five important reasons why gardens, and buildings, are not akin to multiple works such as, for example, woodblock prints. First, it is essential that an originary woodblock (template) is executed and that it continues to exist for as long as new woodblock prints are to be made from it. By contrast, it is not essential that a garden plan is ever made. A garden can come into being via many routes. There may be a detailed plan, but a garden may equally be the result of a rough sketch or verbal instructions, or it may develop in a seemingly haphazard manner in the absence of any clear directional drawn or verbal "plan." Second, once a garden is made it must continue to change, or die. An individual woodblock print generally does not change markedly and, if it does, it may be restored to its original condition. Third, woodblocks relate to the prints derived from them in a close, causal, relationship. No such relationship exists between a garden plan, if one exists, and its built embodiment. Fourth, a woodblock embodies all the features of any prints made from it. At best, a garden plan provides a degree of spatial information. There is no agreed method for a plan to provide the temporal information all gardens require to exist successfully, whether their lifespan be six or 600 seasons. Moreover, because a woodblock embodies all the features of prints made from it, a woodblock may be the source of many near-identical prints. Gardens are not reproducible in this way. They are context dependent and context influencing to a degree unparalleled in the arts.[4] Fifth, Davies claims that an artwork exists if the woodblock itself exists. No instances of it need exist.[5] Similarly, if all prints from a woodblock are destroyed the work survives. The same claims cannot plausibly be made for gardens.

4 See my discussion of gardens' non-reproducibility at 129–33.

5 Davies, "Ontology of Art," § 2.3.

Although I have rejected the idea of gardens being multiple works, there are two respects in which some atypical gardens do in fact share features with them. Gardens designed for show or competition purposes, for which physical, geographical, and climatic contexts are largely immaterial, may be considered performances because such gardens are theoretically able to be "performed" on more than one occasion and in more than one location. They are even able to be "performed" simultaneously on different continents, just as is the case for traditional works for performance.[6] And, scrupulously detailed plans of such gardens may bring to mind the originary templates of non-performance multiple works, such as lithographs and cast sculptures. However, these features do not entail that gardens are performances, or non-performance multiple works. Although show and display gardens' aesthetically relevant features do not usually supervene on physical properties, such as their locations and seasonality, or on their relationships with their geographical and climatic environments and social and historical context, these properties still exist in such gardens, and gardens therefore remain singular works.

There is a further respect in which gardens can be likened to performances: both gardens and performances necessarily exist in a temporal dimension. However, simply existing in a temporal dimension does not necessarily equate to being a performance. Davies writes of non-performance works with a temporal dimension, such as novels and un-performed poetry, that "temporality is not part of their identity *qua* work-instances, because the works they are of do not require in their instances a more or less given duration." By contrast, "it is only works that do require [a more or less given duration] that can be performed ... Performances are events that take place in continuous chunks of real time, where the duration (and separation) of those chunks is a function of the identity of the piece the performance is

6 I am using the terms "performance" and "performer" in the traditional sense of referring to arts such as ballet, music, and theatre and to those who execute works in those art forms. I discuss the pan-art performance concepts developed by Currie and D. Davies in Chapter 6.

of."[7] So, drawing a long bow, it might be said that a spring garden can only be "performed" for a specific number of days each year, after which a specified length of "intermission" follows before its next annual "performance." Similarly, gardens based on nocturnal flowering and scents do not exist in that form during the day, nor bog gardens during times of drought. A visitor cannot jump ahead to the summer garden during winter, although, as I explain later, the direction and duration of the visitor's *experience* of a garden is itself temporally unconstrained.

There is yet another way in which gardens may be likened to performances. This conception was introduced by Salwa in 2013.[8] Unlike D. Davies' performance conception, which was developed in a pan-(performance)art context, Salwa's conception has been developed specifically around the case of gardens. He cautions that his conception is merely one, potentially useful, metaphorical way of regarding gardens. He further cautions that his claims are epistemological; but, as will become clear, they also entail some ontological implications.

Salwa rejects what he sees as the standard comparisons of gardens with paintings, poetry, and architecture because he correctly claims that they undervalue gardens' temporality and dynamism, and the role(s) nature plays in them.[9] He also correctly claims that gardens' nature-like qualities have to be minimized when considered under such non-performance art concepts. Under his garden-performance conception, nature performs in the garden. It contributes freely its "dynamic, changeable and temporal character." Gardens are understood to be "'constant processes,' or 'actions' performed by nature and partially planned by humans."[10] In Salwa's view,

7 Davies, "Ontology of Art," § 2.5. Also see: Levinson and Alperson, "What Is a Temporal Art?," 441, 42, 45.

8 Salwa, "The Garden as a Performance."

9 See ibid., 372, 80–2. I have made preliminary comparisons between gardens and music, and gardens and dance, which attempt to take these factors at least partially into account. See: "We Do Not Have an Adequate Conception of Art until We Have One That Accommodates Gardens."; "Music, Gardens, and Time," in *Time Theories and Music* (Department of Music Studies, Ionian University, Corfu, 2012).

10 "The Garden as a Performance," 381, 83.

gardens combine nature and culture and overcome that (perceived) dichotomy by performing it.[11] Understanding gardens as performances usefully emphasizes their event-like, processual character, and their mutability. It also brings with it five ontological or quasi-ontological consequences.

First, Salwa says that just as in the case of, say, composers and compositions, a garden designer "will never see her piece accomplished, performed once and for all."[12] With regard to performances in general, (S.) Davies finds this view less than convincing but I believe it accommodates well the continuously dynamic nature of gardens, which must embrace events ranging from Dutch elm disease infection, gales, and drought through to the to the subtle introduction into the garden of new turf cultivars.[13] In this way, the appearance of a garden changes continuously until that garden is destroyed. But that does not entail that gardens are performances.[14]

Second, if the garden is understood to be a performance, and visitors to be performers in it, then one can ask with Salwa whether the performance is authentic or, indeed, valid? "Do we and does nature perform in the way it was planned by the architect or gardener?"[15] And this may be a useful critical and experiential question to consider.

Third, performances have a determined beginning and end and, usually, a pre-determined starting and finishing time. By contrast, although a visitor's access to a garden may be limited to opening times and durations, a garden's existence is not usually bound by such constraints.

Fourth, performances not only have beginnings and ends but they also often have internal divisions for artistic reasons (overtures, acts, movements, entr'actes, etc.) and for human reasons (intervals). Performances can

11 Ibid., 373.

12 Ibid., 382.

13 For Davies' view see: Davies, "Ontology of Art," § 2.6.

14 Non-garden works are starting to appear whose temporal extent is extremely long and whose identity is therefore unknowable by any one person. For example, John Cage's *As Slow As Possible* is currently being played on an organ in Halberstadt, Germany, in a performance that will not finish until 2640. More extreme by far is John F. Simon's computer-based visual work *Every Icon*, any completed performance of which is estimated to take "several hundred trillion years."

15 Salwa, "The Garden as a Performance," 384.

stop and start as required and in this way they are unlike gardens, whose processes are necessarily continuous.

Fifth, gardens cannot be performances because they lack performers. It is reasonable to claim that gardens possess some constitutive features in common with performance art forms, such as theatre and dance, because their appeal is in part the result of a viewer's paying attention to objects which are perceived (mainly) visually and which change in various ways over time and in space.[16] For example, in dance a viewer's attention is in part properly directed towards changes in the form, aspect, and location of dancers and their costumes and props. And just so in gardens, a viewer's attention is properly directed to the ways in which living and inanimate objects change, or remain the same, with regard to their form, aspect, or location; or, in Salwa's terms, to how they perform. Performers perform by intentionally interpreting instructions provided in a script, score, or choreography, or they improvise in response to some stimulus provided to them or derived by them from their environment. Plants do not perform. When a gladiolus flowers, or merely exists, it does not do so in response to a designer's instructions: it does so because that is what a gladiolus does at a certain time in its life cycle. While it is true that a garden designer may have located a gladiolus in the knowledge (and hope) that it would produce a red flower on or close to a particular day, even if all goes to plan and the gladiolus flowers on the "correct" day, that does not entail that the gladiolus is performing, because the gladiolus is not interpreting and following the designer's instructions in this, or any other, regard. Plants cannot have intentions and are on that account incapable of performance.[17]

A final distinction can be made between the changing form and aspect of living materials in a garden and the performance arts. In dance, for example, a dancer's being in the early stages of pregnancy, or growing

16 However, the changes in gardens are more frequently multi-sensory than is the case in other performance arts. For example, changing and moving scents are smelt, moving water of a certain temperature is touched, blowing wind is felt, etc.

17 For an interesting contrasting view see Micahel Marder's account of plants performing the unperformable, that is, vegetal growth: Michael Marder, "The Place of Plants: Spatiality, Movement, Growth," *Performance Philosophy* 1 (April 2015).

(imperceptibly) older during the performance, are usually immaterial to the aesthetic content of a ballet, whereas it is precisely such processes of natural change that are at the heart of a garden's aesthetic appeal. But this does not entail that gardens are performances.

An interesting comparison can be made between John Cage's notorious music performance piece *4'33"*, in which whatever ambient sounds occur constitute the performance, and a garden, in which whatever is going on might be said to constitute the "performance." However, in the case of Cage's work, the three constitutive aspects of performances that I have referred to are all put firmly in place by the composer. First, the very title of the work specifies a beginning and end, and results in an exceptionally and unusually tightly defined duration. Second, the piece has an internal sequence of "events," indicated in the score by the instructions for page turning by the performer(s) at pre-determined intervals. Third, detailed instructions are provided to the performer(s) about how to perform the work, including, as is well known, instructions to not make any sound at all. So, while *4'33"* remains controversial as a piece of music, and while it is acknowledged that whatever happens during its duration provides its content, it remains uncontroversially a performance because it fulfils the three important criteria required for the constitution of a performance. The case of *4'33"* reinforces the fact that a garden, even though whatever happens in it may be said to constitute the work, is not a performance because it fails to satisfy the same three criteria.

There is, however, one important way in which gardens are like *4'33"*. I have already remarked on this similarity, but it is worth repeating here because it differentiates gardens not only from other performance works but also from paintings and sculptures; and furthermore, it is constitutive of the nature of gardens. Whatever sounds occur during a performance of *4'33"* and whatever is going on during a visit to a garden constitute the composition or garden on that occasion. Performances of *4'33"* occurring in a soundless anechoic chamber or in a noisy shopping mall sound completely different but still constitute performances of the same work. Similarly, a visit to a flowerless woodland garden on a hot summer day differs radically from a visit to the same garden on a misty day in early spring, when the trees are bare and the camellia flowers glow in the damp air, yet a garden visitor easily accepts that both visits amount to visiting the same garden. However, a visitor's easy acceptance of such differences in the appearance of a garden does not entail that she is viewing a performance. Rather, it means that gardens constitute a uniquely changeable, but still singular, art form.

Gardens Are Singular

One of the most obvious differences between singular and multiple artworks is that the former admit of little change, except for purposes of restoration and conservation, whereas multiple works admit of considerable change over time. It might therefore be expected that gardens are some type of multiple art, but gardens are indeed singular, and they are so despite the degree to which they change, and despite the similarities some may have with some certain multiple work art forms.

Some gardens are relatively a-temporal and stable. For example, many highly formal French Renaissance gardens change very little over time. In these types of gardens, elements that are subject to change are regularly – sometimes even daily – restored to their "original" condition by activities such as raking of grass and gravel and pruning of topiary plants. Such gardens might be considered an equivalent to traditional singular artworks, such as paintings, sculptures, and architecture, which are "restored" by regular or occasional maintenance to what is considered to be a preferred aesthetic state or condition from some prior moment of time.

Ironically, such singular works (gardens) may *appear* to be open to being multiply-instanced because they *appear* to be unchanging and are typically less affected instrumentally and contextually by their physical environment. However, such gardens are no more open to being multiply-instanced than gardens in more changeable styles are. Such gardens do still change, engage with their physical context, and react to climate and weather, although admittedly to a lesser degree than other gardens.[18]

18 It is interesting to note the coincidence, in the case of these realtively unchanging gardens, between gardens' features and gardens' aesthetically relevant properties. These gardens are spatially clearly delineated and contained and often comparatively

However, a-temporal gardens have not been the only model for Western gardens since the Renaissance and throughout this study I have stressed the important and distinctive mutability and temporal nature of many post-Renaissance Western garden styles. These qualities arise principally from the naturalness of living garden elements and from the predictable and unpredictable environmental effects of weather and climate, and seasonal and diurnal cycles. Gardens exist and persist in four dimensions. In experiential terms, they are spatio-temporal entities whose visual, tactile, aural, olfactory, kinaesthetic and, at least in a promissory sense, gustatory stimuli are always changing. Their mutability and temporal nature give rise to ontological complexities which, until most recently, have been, among the arts, unique to the case of gardens.[19]

When a visitor views or otherwise experiences any garden, that garden is at that time always, and inevitably, unlike that garden at any other time in its existence. It is different in countless ways from the "same" garden that existed a minute, day, week, year, or decade ago. Yet it does not usually seem that way to a visitor. In some cases that may be because the differences are slight. But in other cases, such as the visual contrast between visiting an iris garden in summer and then in winter in a cold climate, when the visitor sees no leaves or flowers but plenty of snow, the differences are striking and obvious.[20] How are such complexities to be accounted for? How is this extreme degree of mutability to be explained in the context of the other, unchanging, singular arts?

The answer to this conundrum is that we (unwittingly) accept that gardens are "allowed" to change and retain their singular identity just as is the case for humans, and I suggest that thinking of gardens' ontology in

---

unchanging, *and*, aesthetically relevant properties of these gardens typically did not supervene on physical properties such as location, relationship with environs, seasonality, and mutability.

19 New artforms which have emerged during the last 40 or so years, including land, environmental, installation, and computer art also exhibit somewhat similar, complex ontological characteristics.

20 The degree of mutability exhibited, and accepted, in gardens like the iris garden is unique to gardens and surpasses that of traditional performance arts and the type-token multiple arts.

this way may be more helpful than trying to align gardens with ontologies traditionally associated with other art forms.[21] Gardens change because their principal components are alive and because they are subject to changes wrought on them by environmental and human agencies. In these ways, gardens are importantly like humans. Humans too are living and are altered by environmental and human impacts on them. Just as we readily accept that a person remains the same person despite their being at different times premature, obese, tanned, ill, blond, hirsute, educated, lonely, or an amputee, so too we accept the same garden's being at different times newly planted, badly maintained, bare of leaves, full of flowers, mature, storm damaged, or lacking a view it formerly had. Many of these changes in persons and gardens are the predictable outcome of natural cycles but others are less predictable, or completely unexpected, and result from impacts caused by the environment and humans.

Davies writes that factors pertinent to a work's identity include: "the work's genre, style, and medium, its creator's intentions, and the work's relation to the artist's body of works, to other works to which it refers or by which it is influenced, to the art-historical setting in which it originated, and to the wider social and political environment."[22] I agree with his reasons but would now add, contra Davies, that in the case of gardens, "the work's genre … and medium" entail that its identity is similar, but not identical, to the identity of humans, plants, and animals.

21 I employ this intuitive, non-technical conception of personal identity only as a heuristic device. Within philosophy, the topic continues to be hotly contested. For example, the "worm" theorists believe that a persisting object is composed of the various temporal parts that it has. They believe that all persisting objects are four-dimensional "worms" that stretch across space-time and that you are mistaken in believing that chairs, mountains, and people are simply three-dimensional. By contrast, the "stage" theorists take an object to be identical with a particular temporal part of it at any given time. So, in a manner of speaking, a subject only exists for an instantaneous period of time. For an overview of the topic see: Andre Gallois, "Identity over Time," in *The Stanford Encyclopedia of Philosophy*, ed. Edward N. Zalta (Summer 2012).

22 Davies, *The Philosophy of Art*, 89.

Miller claims that a garden's identity amounts to its geographical location.[23] I contested her claim earlier by proposing that a garden's identity is not just a matter of *where* it is, but of *what* is where it is; and to this proposition I would now add, *when* it is. Gardens, like people, have beginnings and, eventually, endings. It is what happens, continuously, between those two temporal extremes, at a particular location, that constitutes the identity of a garden.[24]

Experience

When theories claiming to be adequate accounts of the experience of the arts are applied to gardens, they prove unsuccessful in accounting for the unique aspects of garden experiences.[25] Such accounts have typically been based on theories developed in the context of other, non-garden arts. They have assumed that gardens function in a restricted range of sensory modalities and that they are distinct, unchanging objects to which we have complete epistemic access from stationary, preferred viewpoints, whereas the actual experience of gardens is shaped by their often indistinct external and internal boundaries, their continuous mutability, and their all-encompassing nature. More importantly, the experience of gardens is shaped by their immersive nature, their multi-sensual, including kinaesthetic, appeal, and by the high degree of free agency a visitor enjoys in "structuring" her garden visit.[26]

23 See: Miller, *The Garden as an Art*, 76–7.

24 Note, however, that the identity of humans and animals is not fixed geographically, although it may be argued that it is usually fixed in the case of plants.

25 For important accounts of the experience of art in non-analytic philosophical traditions see: Berleant, *Art and Engagement*; Dewey, *Art as Experience*.

26 For an interesting art-historical account of the body in eighteenth-century gardens see: Karen Lang, "The Body in the Garden," in *Landscapes of Memory and Experience*, ed. Jan Birksted (London: Spon Press, 2000).

Although in some formal gardens, for example those at the palace at Versailles, where Le Nôtre appears to have laid out parts of the garden so that they are best viewed from the privileged perspective offered from the king's quarters, and although, with reference to the eighteenth-century circuit garden, Ross writes that it is important to remember the degree to which such gardens "manipulated and controlled their visitors' experiences," most gardens are not controlling in this way.[27] Most gardens allow, and garden makers expect, visitors to explore in a spatially and temporally undirected manner. Hunt writes that "nowhere in any fine garden is the visitor permitted an adequate view of the whole – the process of understanding even the smallest territory and its changes through hour and season militate against that."[28]

Birksted articulates clearly the fundamental and distinctive features of the experience of gardens. Preferring the multisensory "beholder" to the merely visual "visitor," he writes of the "spatial and temporal location of the (extra)visual and mobile beholder in the three dimensions of ... gardens," and of the beholder's "variable perceptual structures."[29] He separates out "focalization," the garden maker's point of view, from perspective, or the beholder's point of view, and argues that they are inextricably linked in an adequate garden experience. And he notes that "the importance of temporality [in the experience of gardens] forces one to reconsider the centrality

27 Ross, *What Gardens Mean*, 159.

28 Hunt, *Greater Perfections: The Practice of Garden Theory*, 7. However, Herrington points out that some gardens, for example Schwartz's *Bagel Garden*, are to be looked at, not moved through. (Susan Herrington, *On Landscapes*, eds Simon Crichtley and Richard Kearney, Thinking in Action (New York; Abingdon, Oxon: Routledge, 2009), 3–4.) But such gardens remain three-dimensional works and a viewer may move her viewpoint and thereby see them differently, or, she may imaginatively enter and move about in them.

29 Birksted, "Landscape History and Theory: From Subject Matter to Analytic Tool," 7. It is perhaps just an etymological quirk, but it is none the less interesting to note that although we attend a concert or performance, read a book, look at a painting, and so on, we *visit* a garden. "Visit" has its etymological roots in the Latin *visere*, which means "to see", and perhaps this unwittingly contributes to the tendency to emphasize the visual above all else in conceptualizing gardens.

of space as the predominant dimension, and highlights the complex interactions between space and time."[30]

The understanding of the garden viewer as mobile in space, and therefore in time, is not new. In the first half of the eighteenth century, Walpole was already acknowledging such when he wrote that the "animate prospect is the most continually rewarding garden feature."[31] However, this essential feature of the experience of gardens has remained largely unexplored, one might say almost ignored, by philosophical aesthetics.

A garden is always changing and, in any but the smallest garden, a viewer's experience of a garden is inevitably partial. A visitor can know only one "time slice" and one, or a series of, "space slices" of a garden at a time. This means that direct epistemic access to a garden is doubly restricted, and it may be speculated that this is similar to what occurs when one person experiences another person. In the case of persons, each has direct epistemic access only to the time/space slice of the person that is existing in front of her. Just as in the case of gardens, we "construct" people from different bits and pieces of information from now, the future, and past, we potentially use all our senses, and, apart from some limited, atypical situations, there is no recommended or preferred viewpoint, procedure, or direction to follow in doing so. Therefore, it might be proposed that the garden itself is, in the senses just described, person-like.

If we conceptualize the garden in this way, then the *experience* of a garden will necessarily be unlike the experience of works in other traditional artforms. There is an account of another art that might usefully be employed. It is a phenomenologically based account of contact improvisational dance, an account that helpfully illuminates what it is like to be a person experiencing a garden.[32]

30 Ibid., 11.

31 Quoted in: Ross, *What Gardens Mean*, 163.

32 This account does not entail that the garden is a performance, which is something I have been at pains to deny. I invoke the account simply to improve the understanding of what being a garden *visitor* appropriately entails, and how it differs from the experiences of other artforms.

The philosopher Philip Alperson writes of musical improvisation but it is reasonable to take his account as being applicable to other forms of artistic improvisation also.[33] He writes:

> The chief characteristic of improvised activity, it seems to me, is that while we might or might not have a general idea of the sort of outcome of the activity in question, in improvised activity, certain of the fundamental features of both the activity and of the product of that activity are determined in the very doing of the activity. In a very general sense, we can think of improvisation as a kind of goal-directed activity, ... but what makes the activity improvisatory is the sense that what is being done is being done on the fly.
>
> [However,] if we think of improvisational activity along these lines, there would [be] an indefinitely large number of human activities that involve improvisation. Indeed, improvisation would seem to be a feature of most, if not all, directed human thought and action. ... So, what makes art improvisation different? ... We designate activities as improvisatory, I think, not simply in cases where we notice an evident degree of spontaneity or free play in the activity. Improvisation is not completely free or autonomous activity. Improvisation depends fundamentally on routines, rituals, and practiced activities of all sorts.[34]

Alperson's account of musical improvisation neatly accommodates many of the behaviours of garden visitors. Just as in the case of musical (jazz) improvisation, where a musician, or group of musicians improvise on some pre-existing musical material, so too in gardens, a visitor, or group of visitors improvises on, around, and over the garden base supplied by the garden designer.[35] Matters of sensory involvement and precedence, and spatial,

33 Philip Alperson, "A Topography of Improvisation," *The Journal of Aesthetics and Art Criticism* 68, no. 3 (2010).

34 Ibid., 273.

35 This idea is not unrelated to the pragmatist Dewey's conception of the work of art: "'The product of art – temple, painting, statue, poem – is not the work of art. The work of art takes place when a human being cooperates with the product so that the outcome is an experience.'" (Quoted in: Susan Herrington, "When Art Is a Garden: Benny Farm by Claude Cormier," in *Contemporary Garden Aesthetics, Creations and Interpretations*, ed. M. Conan (Washington, DC: Harvard Univeristy Press, 2007), 20.)

temporal, and locomotional engagement and preferences are "determined in the very doing of the activity."[36] Just as in the case of musical improvisation, a garden visit is "not [a] completely free or autonomous activity. [It] depends fundamentally on routines, rituals, and practiced activities of all sorts."[37] And, garden visits need not be solitary affairs; a visiting group may have a social dimension encompassing "conversational" and social protocols, just as in musical improvisation.

The improvisational account of the experience of gardens remedies two of the objections raised earlier to Salwa's performative account. First, unlike performances, whose length is determined within certain limits, improvisations are of uncertain length (and gardens of potentially indeterminate length). Furthermore, improvisations may feature un-programmed internal temporal divisions. Second, improvisations are by nature unrepeatable – if repeated, they would become *performances* of "improvisations" – and, because gardens change perpetually, garden experiences are in this sense unrepeatable. Furthermore, neither improvisations nor garden experiences necessarily entail the presence of an audience.

Successfully comparing gardens to improvisations has a significant hurdle yet to overcome. Alperson writes of the high-level musical and technical skills that competent improvising demands.[38] While there is no denying that an adequate sensory perceptual apparatus is important and a knowledge of gardening and its traditions advantageous to a visitor, a high level of (performing) skills is unnecessary for the garden visitor. A way around this apparent impasse is to be found in the theories and practices of the improvisatory dance that developed in USA from the 1970s onwards and whose repercussions continue to resonate internationally.

Writing of improvisation in dance, the philosopher Curtis Carter describes how, during the twentieth century, dance and the other arts have "undergone major changes reflecting increased democratization and open form"; how the "mainly hierarchical systems of earlier centuries"

36 Alperson, "A Topography of Improvisation," 273.

37 Ibid.

38 Ibid., 275–7.

have been replaced by "collective participatory practices."[39] In what follows I will, unless otherwise noted, be drawing on his account from 2000, "Improvisation in Dance." Carter writes:

> Improvisation invites examining a situation from various angles that can be invented in the very process of creation; [that] the primary instrument through which improvisation in dance takes place is the human body and its interactions with other bodies [and the environment (see p. 178)]; [that] the full range of human attributes, including the physical, conceptual, and emotional resources embodied in the body are thus available for improvisation in dance; [and, finally, that] richness and variety are brought to the improvisation process through the aid of immediate feedback, which could completely change the direction of events. ... The improviser must create the artistic product as he or she performs it.[40]

All of these quotations from Carter mirror well the account of the improvisatory model of the experience of gardens. Carter then goes on to describe the contact improvisation introduced by Steve Paxton and others in which three additional aspects of the new style of improvisation are commented on. Each of these aspects is important for the improvisatory model of the experience of gardens.

First, discussing the style of the influential choreographer Yvonne Rainer, Carter notes that it was, and still is, often characterized by an interest in "new uses for everyday movements of people without formal dance training."[41] Furthermore, the temporal sequence of such improvised movements is to be "determined" during the course of the improvisation by the (non-)dancers. Thus, in the garden context, garden-improvisers need not

39 Curtis L. Carter, "Improvisation in Dance," *The Journal of Aesthetics and Art Criticism* 58, no. 2 (2000), 181.

40 Ibid., 181–2.

41 See: ibid., 186. Carter refers in particular to: Yvonne Rainer, *Work 1961–73* (Halifax, Nova Scotia: Press of the Nova Scotia College of Art and Design, 1974). Danto commented of Rainer that she "grasped the Sixties question of distinguishing merely sitting in a chair and a dance movement consisting of sitting in a chair." See his "Ontology, Criticism, and the Riddle of Art Versus Non-Art in *the Transfiguration of the Commonplace*," *Contemporary Aesthetics* 6 (2008), § 7.

be trained or skilled in their art and nor need their garden-improvisation be structured or pre-arranged in any way.[42]

Second, in his discussion of contact choreographer Paxton's approach to improvisation, Carter writes that what matters is a body's physical, sensory, and emotional interactions with the environment, including other bodies.[43] He summarizes anthropologist Cynthia Novack's view of contact improvisation as, "inward focussed, informal in dress and manner, involving pedestrian movement."[44] Thus, in the garden context, garden-improvisers' bodies react with their environment in ordinary, non-specialized ways.[45]

Novack hints at the third important aspect of contact improvisation when she describes it as "inward-focussed." Carter takes this further: "In its inward-turning mode, contact improvisation changed the relationship of performers to audience. Instead of performing to or for an audience, the dancers in contact improvisation are focussed of the action occurring among themselves."[46] Thus, although we might expect an improviser to have an audience, that is not so in the case of contact improvisations. A garden visitor "improvises" her encounter with the garden for her own benefit, not for any audience, and the conception further allows that a garden visitor may "improvise" her visit jointly with other garden visitors, thereby accommodating a feature of some garden experiences.

42 This attitude to the involvement of (non-)dancers was not new. Carter writes that "Laban[, working in Zurich at the time of the Dadaists,] extended his theories on improvisation to the training of amateurs and thus anticipates the blurring of the lines found in post-modern dance between the movements of ordinary persons and trained dancers in performance." See Carter, "Improvisation in Dance," 182.

43 Ibid., 186. For an explanatory introduction to Paxton's work see his "Contact Improvisation," *The Drama Review* 19 (1975).

44 "Improvisation in Dance," 187. For a fuller exposition of Novack's views on these aspects of contact improvisation see: Cynthia J. Novack, *Sharing the Dance: Contact Improvisation and American Culture* (Madison: University of Wisconsin Press, 1990), 128–9.

45 Dewey claims that "bodily engagement and resistance are central to the development of the process" (*Art as Experience*, 214). This is different from what I am getting at here because without bodily engagement garden appreciation, as opposed to, say, reading a novel or contemplating a painting, is largely impossible.

46 Carter, "Improvisation in Dance," 187.

This new conception of dance raised profound questions for that performance-based art form: How are "performances" to be evaluated? How are they to be repeated? What role is a "performance" audience to have? How is that audience to be engaged and satisfied? And, once the proscenium theatre had been abandoned, what role is any performing environment to play in the "performance?" These questions, once transferred to the experience of the art form of gardens, become so familiar as to be unremarkable. They are questions with which garden art has comfortably lived for centuries. Gardens have always been experienced in these ways. Contact improvisatory dance has unknowingly taken on these aspects of the garden experience and in so doing has usefully provided an explanatory framework for conceptualizing the garden experience as an improvisation.

Although the explanatory usefulness of the metaphor of contact improvisation is considerable, garden visiting, and improvisational dancing remain dissimilar in a significant way: the participants in the former are not intending to, or deliberately participating in, an improvised performance, whereas participants in the latter are. This dissimilarity does not weaken the explanatory usefulness of the metaphor because my aim, in employing it, is to characterize the activities themselves, rather than the state(s) of mind of the participants. Garden visitors are intending to visit a garden and contact dancers are intending to dance. My interest has been in examining and finding similarities between the outcomes (activities) stemming from each party's intentions.

Gardens' Materials

Gardens are like Brillo boxes.[47] We are so familiar with plants, paths, water, wind, sun, soil, birds, and almost all the other familiar constituents of gardens that, like Brillo boxes, but unlike *Brillo Boxes (1964)*, we take them for

47 See footnote 10, p. 144.

granted.[48] Those who gathered at the Stable Gallery in New York for the 1964 opening of *Brillo Boxes* were suitably troubled by that work's seeming ordinariness and its mundane, quotidian, not to say non-art, aspect. I claim that gardens "deserve" a similar reaction because their contents also comprise ordinary, mundane, and quotidian objects. There is scarcely anything in a conventional garden that does not exist as an *objet*, or *évènement trouvé*, somewhere outside the garden. Furthermore, garden makers organize and present their quotidian objects for our appreciation, just as Andy Warhol did. However, unlike the viewers at that Stable Gallery opening in 1964, garden visitors are inured to the ordinariness of gardens' materials – they even find many of them intrinsically attractive – and this in part explains why gardens continue to slip under the philosophical radar or, rather, why there appears to be nothing problematic about them that warrants philosophical investigation. Danto neatly expressed an opinion with regard to (presumably) baffling, "anxious objects": "the objects approach zero as their theory approaches infinity."[49] And so, as gardens are in this sense far from "approaching zero," their philosophical theorizing has consequentially fallen behind that of other, more problematic arts.

Not only gardens' materials but also their spaces and temporality are ordinary. Garden spaces often appear indistinguishable from real-world spaces, and shade, for example, is as welcome in a garden on a hot day as it is outside the garden, and it takes the same time for a plant to grow or for a visitor to age inside and outside a garden. I have earlier rejected Miller and Ross accounts of gardens' virtuality and I would now add this further,

48 Danto wrote of *Brillo Boxes* that "it asked, in effect, why it was art when something just like it was not." (Arthur Danto, "Approaching the End of Art," in *Richard Rudner Memorial Symposium* (Washington University, Saint Louis, MO, 1987), 24.) The question is not new. Quatremère de Quincy, an astute (according to Turner) French critic, wrote of the landscape garden in 1823: "What pretends to be an image of nature is nothing more or less that nature herself. The means of the art are reality. Everyone knows that the merit of its works consists in obviating any suspicion of art." See: Turner, *Garden History: Philosophy and Design 2000 B.C.–2000 A.D.*, 226.

49 Arthur Danto, *The Philosophical Disenfranchisment of Art* (New York: Columbia University Press, 1986), 111.

straightforward reason: gardens' unmistakable and unavoidable realness and aliveness make them unlikely candidates for virtuality.[50]

Ross invokes theories of Wollheim, Wolterstorff, and Danto to support her view of gardens as virtual entities. She explains Wollheim's theory of twofoldness this way: "In viewing a representational painting we are simultaneously aware of both the painted surface and the image seen in that surface."[51] However, even so-called representational gardens are only partially so, and many gardens are, *pace* Hunt, not representational at all. Therefore, Ross's claim that a physical (non-representational) garden and its virtual counterpart function in the way Wollheim describes for representational paintings does not succeed because "nothing," that is, no image, is necessarily being intentionally represented by the garden designer's choice and arrangement of her materials. Moreover, I believe that the linking of, in Wollheim's case, an unchanging painted image with, in Ross's case, a constantly changing garden "image," is misleading and further weakens Ross's claim.

Ross also invokes Wolterstorff's "world of the work" to support her claims regarding the simultaneously actual and virtual worlds of the garden. Wolterstorff's arguments originated in the literary (narrative) arts and for this reason, just as in the case of Wollheim's theories, they were developed in response to unchanging works, though not experiences, of art. Wolterstorff's account of an author's "indication" and that author's readers' subsequent "elucidation" and "extrapolation" in the creation and experience of literary works is applicable in its extensiveness to the case of gardens, and Ross expands on this account. However, overall, in applying Wolterstorff's theories, Ross restricts herself almost completely to an account of gardens' visual effects, and, further, to effects that might be experienced by a static viewer at a given time and from a given viewpoint, thereby denying gardens the multi-modality and multi-sensuality she describes and celebrates elsewhere.

Elsewhere, Ross invokes Danto's theories to explain how "ordinary" objects – like hedges and lawns (or Brillo boxes) – can function as virtual

50 See 86–88.
51 Ross, *What Gardens Mean*, 178.

art objects.[52] Danto's explanation of why a necktie daubed by a child is non-art whereas an identical necktie resulting from Picasso's daubing may be a work of art is well known. In essence, the necktie daubed by Picasso is a work of art because it invites or requires interpretation, and this "consists [in part] in determining the relationship between ... [the] work of art and its material counterpart."[53] Using this argument, Ross distinguishes the garden work of art from its physical, or material, counterpart. However, this does not necessarily entail, as it does for Ross, that the garden becomes a virtual entity. I believe that it means no more than that the garden requires interpretation, that its *esse* is *interpretari*, that, in Danto's words, the artworld is "a world of interpreted things."[54] Gardens are usually more attractive to the senses than Brillo boxes and, in some ways, this makes them harder to conceptualize as art on Danto's conception. We usually do not "need" to separate the art garden from the physical garden in order to enjoy it, in the way that the bewildered viewers did in the case of *Brillo Boxes*. But, philosophically, the tasks and the reasons for them are similar.

Ross's use of Danto's theory is right as far as it goes but the reality is more complex than she acknowledges in her discussion, although she refers positively to those same "complexities" elsewhere in her book.[55] Gardens are unlike *Brillo Boxes* because their "material counterpart" is itself

52 Ibid., 180–6.

53 Ibid., 179.

54 Danto, *The Transfiguration of the Commonplace: A Philosophy of Art*, 113. Danto was not alone in promoting an understanding of the use of mundane, "obvious," or clichéd materials in art. For example, the Russian theorist Vikto Shklovsky developed a psychological theory, *ostranenie*, in the 1920s. *Ostranenie* has been applied to the case of a contemporary garden by landscape architect Jacky Bowring. See: Jacky Bowring, "'To Make the Stone[S] Stony': Defamiliarization and Andy Goldsworthy's Garden of Stones," in *Contemporary Garden Aesthetics, Creations and Interpretations*, ed. M. Conan (Washington, DC: Harvard University Press, 2007).

55 I refer here particularly to gardens' multi-sensual, multi-modal, and multi-dimensional qualities. Not only is it necessary to "see as," to employ Wittgenstein's phrase, but it is equally necessary to smell as, feel as, hear as, move as, and so on. For a good, fresh account of the complex, inter-related elements and experiences of gardens and landscapes see: Herrington, *On Landscapes*, 3–16.

constantly changing in a multitude of ways and modes. Furthermore, the fact that the garden beholder is "(extra)visual and mobile … in gardens' three-dimensional space and time" adds a complexity (understandably) ignored by Danto in his theory.[56] In the case of the garden, not only is the material counterpart continuously changing of its own accord but also the beholder's multi-modal experience of it is so highly idiosyncratic and variable that the material counterpart of the garden may be said to be changing, or at least reconstructed frequently, on those grounds too.

If Danto's theory of art can be understood to accommodate these two levels of continuous change and the varying modalities in the material counterpart of the garden work of art then it is, of all the theories re-presented in the literature, the most apposite to the case of art gardens. Such an understanding of Danto's theory would account not only for the ordinary, real, fictive space of gardens, as Birksted required an adequate theory to do, but also for their ordinary, real, fictive temporality.[57] It would allow us to counter Roger Scruton's claim, that a tree grows outside a garden but performs its growth inside a garden, with the observation that the growing is not different within or outside the garden; it's just that in the latter case we ought to be interpreting that growing.[58] It would allow us to understand how the various types of ordinary time and temporal processes Miller cites as obtaining in a garden – for example, diurnal, seasonal, geological, meteorological, and chronological – could be "transfigured" into artistic times and processes. It would allow us to retain our instinctive awareness of the extraordinary complex blend of sensory inputs gardens offer instead of having to sacrifice that awareness in the face of art theories inimical to the reality of gardens.

There is a further, final way in which Danto's theory is appropriate to the case of gardens. With the exception of Danto's, all the theories discussed above share a significant underlying problem: they each require that we

56 The quoted description of the "beholder" is from Birksted, "Landscape History and Theory: From Subject Matter to Analytic Tool," 6.

57 The useful term "fictive space" is taken from: ibid., 17.

58 See Roger Scruton, *Perictione in Colophon: Reflections on the Aesthetic Way of Life* (South Bend, IN: St Augustine's Press, 2000), 83.

understand gardens as "structures" with "formal" qualities. But, as Danto explained with reference to *Brillo Boxes*, when "an artwork is perceptually indiscernible from an object which is not an artwork ... then form alone neither makes an artwork nor gives it whatever value it has."[59] So, in the case of gardens, whose materials fall squarely into this category of objects, invoking formalist, ontological and epistemological explanations will generally be unsuccessful. It may be said that on Danto's account, the ways in which we (should) understand gardens are similar to the ways in which we understand people. For example, one person's hand is generally like billions of other hands. To the degree that we are interested in that hand, and the rest of the person, we are typically interested in it because it is *that* person's hand, not because it exhibits the structure and form of "hand." The "materials" that make up humans are ordinary and omnipresent. But the activity of knowing and understanding one particular person's materials involves *interpreting* the unique combination and context of those omnipresent, "ordinary" materials, and the interrelationships between them, just as is the case for knowing and understanding gardens. In this way, too, gardens may be thought to be more like persons than they are like works in other, traditional art forms.

59 Quoted in: James Shelley, "The Concept of the Aesthetic," ed. Edward N. Zalta, Winter 2015 edn, *The Stanford Encylopedia of Philosophy* (Winter 2015), <http://plato.stanford.edu/archives/win2015/entries/aesthetic-concept>.

Bibliography

"Aesthetic, N. and Adj.". In *OED Online*, Oxford University Press, June 2016. <http://www.oed.com/view/Entry/3237>.

Alperson, Philip. "A Topography of Improvisation." *The Journal of Aesthetics and Art Criticism* 68, no. 3 (2010), 273–80. doi:10.1111/j.1540-6245.2010.01418.x.

Aristotle. "Poetics."

Assunto, Rosario. "I Giardini Della Parola E La Parola Dei Giardini." *Artibus et Historiae* 3, no. 5 (1982), 17–29. <http://www.jstor.org/stable/1483141>.

Barwell, Ismay, and John Powell. "Gardens, Music, and Time." In *Gardening: Cultivating Wisdom*, 136–47. Oxford: Wiley-Blackwell, 2010.

Beardsley, Monroe C. *Aesthetics: Problems in the Philosophy of Criticism*. 2nd edn. Indianapolis, IN: Hackett Publishing Company, 1981.

Bell, Clive. "Significant Form (Extracts from Chapter One and Chapter Three of *Art*)." In *Aesthetics: Twentieth Century Readings*, ed. E. Hirst. Boston, MA: Beacon Books, 1986.

Benney, Elizabeth M. "Tupare." In *The New Zealander's Garden*, eds Julian Matthews and Gordon Collier, 24–9. Auckland, NZ: Endeavour Press, 1985.

Berleant, A., and A. Carlson. "Introduction." In *The Aesthetics of Human Environments*, eds A. Berleant and A. Carlson. Peterborough, Ontario: Broadview Press, 2007.

Berleant, Arnold. "The Aesthetics of Art and Nature." In *The Aesthetics of Natural Environments*, eds A. Carlson and A. Berleant. Peterborough, Ontario: Broadview Press, 2004.

——. *Art and Engagement*. Philadelphia, PA: Temple University Press, 1991.

——. "What Is Aesthetic Engagement." *Contemporary Aesthetics* 11 (2013). <http://www.contempaesthetics.org/newvolume/pages/article.php?articleID=684>.

Berlin, Isaiah. *The Crooked Timber of Humanity: Chapters in the History of Ideas*. London: John Murray, 1990.

Bertram, Christopher. "Jean Jacques Rousseau." In *The Stanford Encyclopedia of Philosophy*, ed. E. N. Zalta. Winter 2012. <http://plato.stanford.edu/archives/win2012/entries/rousseau/>.

Birksted, Jan Kenneth. "Landscape History and Theory: From Subject Matter to Analytic Tool." *Landscape Review* 8, no. 2 (2003), 4–28.

Bowring, Jacky. "'To Make the Stone[S] Stony': Defamiliarization and Andy Goldsworthy's Garden of Stones." In *Contemporary Garden Aesthetics, Creations and Interpretations*, ed. M. Conan. Washington, DC: Harvard University Press, 2007.

Brady, E. *Aesthetics of the Natural Environment*. Tuscaloosa: University of Alabama Press, 2003.

Brown, David, and Tom Williamson. *Lancelot Brown and the Capability Men: Landscape Revolution in Eighteenth-Century England*. London: Reaktion, 2016.

Burnham, Douglas. "Immanuel Kant: Aesthetics." In *The Internet Encyclopedia of Philosophy*, ed. James Fieser. 2014. <http://www.iep.utm.edu/>.

Carlson, A., and A. Berleant, eds. *The Aesthetics of Human Environments*. Peterborough, Ontario: Broadview Press, 2007.

Carlson, Allen. "Environmental Aesthetics." In *The Stanford Encyclopedia of Philosophy*, ed. Edward N. Zalta. Summer 2012. <http://plato.stanford.edu/archives/spr2015/entries/environmental-aesthetics/>.

Carroll, Noël. "Historical Narrations and the Philosophy of Art." *Journal of Aesthetics & Art Criticism* 51 (1993).

Carter, Curtis L. "Improvisation in Dance." *The Journal of Aesthetics and Art Criticism* 58, no. 2 (2000), 181–90.

Charlesworth, Michael. "Derek Jarman's Garden at Prospect Cottage, Dungeness, and His Avebury Paintings." *Studies in the History of Gardens & Designed Landscapes: An International Quarterly* 35, no. 2 (2015), 172–82. doi:10.1080/14601176.2014.980066.

Clark, Kenneth. *Landscape into Art*. London: J. Murray, 1976.

Conan, M. "Introduction: In Defiance of the Institutional Art World." In *Contemporary Garden Aesthetics, Creations and Interpretations*, ed. M. Conan. Washington, DC: Harvard University Press, 2007.

Coomaraswamy, Ananda. "The Philosophy of Mediaeval and Oriental Art." In *Coomaraswamy*, ed. Roger Lipsey. Princeton, NJ and Surrey, UK: Princeton University Press, 1977.

Cooper, David E. "Foreword." In *Gardening: Cultivating Wisdom. Philosophy for Everyone*, ed. Dan O'Brien, 2010.

——. "In Praise of Gardens." *British Journal of Aesthetics* 45, no. 2 (2003).

——. *A Philosophy of Gardens*. Oxford: Oxford University Press, 2006.

Cowell, F. R. *The Garden as a Fine Art: From Antiquity to Modern Times*. London: Weidenfield and Nicholson, 1978.

Cox, Oliver. "A Mistaken Iconography? Eighteenth-Century Visitor Accounts of Stourhead." *Garden History* 40, no. 1 (2012).

Currie, Gregory. *An Ontology of Art*. New York: St Martin's Press, 1989.

Danto, Arthur. "Approaching the End of Art." In *Richard Rudner Memorial Symposium*. Washington University, Saint Louis, MO, 1987.

——. *The Philosophical Disenfranchisment of Art*. New York: Columbia University Press, 1986.

Danto, Arthur C. "Ontology, Criticism, and the Riddle of Art Versus Non-Art in *The Transfiguration of the Commonplace*." *Contemporary Aesthetics* 6 (2008).

——. *The Transfiguration of the Commonplace: A Philosophy of Art*. Cambridge, MA and London: Harvard University Press, 1981.

Davies, David. *Art as Performance*. Malden, MA; Oxford: Blackwell Publishing, 2004.

Davies, Stephen. "Essential Distinctions for Art Theorists." In *Philosophical Perspectives on Art*. Oxford: Oxford University Press, 2010.

——. "Ontology of Art." In *The Oxford Handbook of Aesthetics*, ed. Jerrold Levinson. New York: Oxford University Press, 2009. <http://www.oxfordhandbooks.com/view/10.1093/oxfordhb/.../oxfordhb-9780199279456>.

——. *Philosophical Perspectives on Art*. Oxford: Oxford University Press, 2010.

——. *The Philosophy of Art*. Malden, MA: Blackwell, 2006.

Dewey, J. *Art as Experience*. New York: Putnam Capricorn, 1958.

Dickie, George. *Art and the Aesthetic*. Ithaca, NY: Cornell University Press, 1974.

——. *The Art Circle: A Theory of Art*. New York: Haven Publications, 1984.

"Dumbarton Oaks Garden and Landscape Studies Symposium: Sound and Scent in the Garden." <http://www.doaks.org/research/garden-landscape/scholarly-activities/past/sound-and-scent-in-the-garden/abstracts>.

Eliasson, Olafur. "Your Rainbow Panorama." <http://en.aros.dk/visit-aros/the-collection/your-rainbow-panorama/>.

Eliot, T. S. "Tradition and the Individual Talent (1919)." In *Selected Prose of T. S. Eliot*, ed. Frank Kermode, 37–44. London: Faber and Faber, 1975.

Elkins, J. "On the Conceptual Analysis of Gardens." *Journal of Garden History* 13, no. 4 (1993), 189–98.

Feagin, Susan L., and Patrick Maynard, eds. *Aesthetics*, Oxford Readers. Oxford: Oxford University Press, 1997.

Ferrari, G. R. F. "The Meaninglessness of Gardens." *The Journal of Aesthetics and Art Criticism* 68, no. 1 (2010), 10.

Fleming, Laurence, and Alan Gore. *The English Garden*. London: Spring Books, 1979.

Fowler, Michael. "Sound as a Considered Design Parameter in the Japanese Garden." *Studies in the History of Gardens & Designed Landscapes: An International Quarterly* (2015), 1–16.

——. "Soundscape as a Design Strategy for Landscape Architectural Praxis." *Design Studies* 34, no. 1 (2012), 111–28.

Gallois, Andre. "Identity over Time." In *The Stanford Encyclopedia of Philosophy*, ed. Edward N. Zalta, Summer 2012.

"Gardenesque, Adj.". In *OED Online*, Oxford University Press, June 2015. <http://www.oed.com/view/Entry/76731>.

Gaut, B. "'Art' as a Cluster Concept." In *Theories of Art Today*, ed. N. Carroll. Madison: University of Wisconsin Press, 2000.

Gillette, Jane. "Can Gardens Mean?" In *Meaning in Landscape Architecture and Gardens: Four Essays; Four Commentaries*, ed. Marc Treib. New York: Routledge, 2011.

Gilpin, William. *Three Essays: On Picturesque Beauty; on Picturesque Travel; and on Landscape Sketching: To Which Is Added a Poem on Landscape Painting*. [Electronic resource.] London: R. Blamire, 1792.

Graham, Gordon. "Expressivism: Croce and Collingwood," in *The Routledge Companion to Aesthetics*, eds B. Gaut and D. McIver Lopes. Abingdon, Oxon: Routledge, 2005.

Hegel, G. W. F. *Aesthetics. Lectures on Fine Art*. Trans. T. M. Knox. Oxford: Clarendon Press, 1975.

Hepburn, R. W. "Contemporary Aesthetics and the Neglect of Natural Beauty." In *British Analytical Philosophy*, eds Bernard Williams and Alan Montefiore. London: Routledge and Kegan Paul, 1966.

Herrington, Susan. *On Landscapes*. Thinking in Action. Eds Simon Crichtley and Richard Kearney. New York; Abingdon, Oxon.: Routledge, 2009.

——. "When Art Is a Garden: Benny Farm by Claude Cormier." In *Contemporary Garden Aesthetics, Creations and Interpretations*, ed. M. Conan. Washington, DC: Harvard Univeristy Press, 2007.

Heyde, Steven. "The Historical Roots of 'Aesthetics' in Landscape Architecture: An Introduction." *Studies in the History of Gardens & Designed Landscapes*. 34, no. 2 (2014).

Hick, Darren Hudson. *Introducing Aesthetics and the Philosophy of Art*. London, New York: Continuum, 2012.

Houlgate, Stephen. "Hegel's Aesthetics." In *The Stanford Encyclopedia of Philosophy*, ed. Edward N. Zalta. Spring 2014 edn. <http://plato.stanford.edu/archives/spr2014/entries/hegel-aesthetics/>.

Hunt, John Dixon. "Beyond Ekphrasis, Beyond Sight, Beyond Words. ..." <http://www.doaks.org/research/garden-landscape/scholarly-activities/past/sound-and-scent-in-the-garden/abstracts>.

——. "Emblem and Expression in the Eighteenth-Century Landscape Garden." *Eighteenth-Century Studies* 4, no. 3 (1971), 304.

——. *The Figure in the Landscape*. Baltimore, MD and London: Johns Hopkins University Press, 1976.

——. "Gardens: Historical Overview." In *Encyclopedia of Aesthetics*, ed. Michael Kelly. New York, Oxford: Oxford University Press, 1998.

——. *Greater Perfections: The Practice of Garden Theory*. Philadelphia: University of Pennsylvania Press, 2000.

——. "Stourhead Revisited & the Pursuit of Meaning in Gardens." *Studies in the History of Gardens & Designed Landscapes* 26, no. 4 (2006), 328–41.

——. "The Time of Walking." *Studies in the History of Gardens & Designed Landscapes: An International Quarterly* 36, no. 4 (2016).

Irvin, Sherri. "Scratching an Itch." *Journal of Aesthetics & Art Criticism* 66, no. 1 (2008), 25–35.

"Jane Kahan Gallery." <http://janekahan.com/tapestries/>.

Kant, Immanuel. *Critique of Judgement*. Trans. J. C. Meredith. New York: Oxford University Press, 2007.

Kienast, Dieter, and Günther Vogt. „Die Form, Der Inhalt Und Die Zeit = Form, Content and Time." *Topos: European landscape magazine*, no. 2 (1993), 6.

Kivy, Peter. *Introduction to a Philosophy of Music*. Oxford: Clarendon Press, 2002.

——. *The Performance of Reading: An Essay in the Philosophy of Literature*. New Directions in Aesthetics. Oxford: Blackwell, 2006.

——. *Philosophies of Art: An Essay in Differences*. Cambridge: Cambridge University Press, 1997.

Knight, Richard Payne. *The Landscape: A Didactic Poem in Three Books : Addressed to Uvedale Price, Esq*. [Electronic resource.] London, 1795.

Kristeller, Oskar. "The Modern System of the Arts." In *Renaissance Thought 2: Papers on Humanism and the Arts*. New York: Harper and Row, 1965.

Lang, Karen. "The Body in the Garden." In *Landscapes of Memory and Experience*, ed. Jan Birksted. London: Spon Press, 2000.

Langer, S. K. *Feeling and Form: A Theory of Art*. London: Routledge and Kegan Paul, 1953.

Leddy, Thomas. *The Extraordinary in the Ordinary: The Aesthetics of Everyday Life*. Peterborough, Ontario: Broadview Press, 2012.

——. "Gardens in an Expanded Field." *British Journal of Aesthetics* 28, no. 4 (1988).

Lee, Rensselaer W. *Ut Pictora Poesis: The Humanistic Study of Painting*. New York: Norton, 1967.

Levinson, Jerrold. "Work and Oeuvre." In *The Pleasures of Aesthetics: Philosophical Essays*. Ithaca, NY: Cornell University Press, 1996.

Levinson, Jerrold, and Philip Alperson. "What Is a Temporal Art?" *Midwest Studies in Philosophy* 16 (1991), 439–50.

Lippard, Lucy R. "Gardens: Some Metaphors for a Public Art." *Art in America* 69, no. 11 (1981).

Lopes, Dominic McIver. "Nobody Needs a Theory of Art." *The Journal of Philosophy* 105, no. 3 (2008), 109–27.

Loudon, J. C. . *The Suburban Gardener, and Villa Companion*. [Electronic resource.] London: The author, 1838. <https://archive.org/details/suburbangardene00loudgoog>.

"A Lyrical Landscape: Toronto's Waterfront Music Garden Bridges All the Senses." *Country Living Gardener*, January-February 2003, 72.

Macarthur, John. "Architecture and the System of the Arts; or, Kant on Landscape Gardening." In *Architecture, Disciplinarity, and the Arts*, eds Andew Leach and John Macarthur. Ghent: A. & S. Books, 2009.

Marder, Michael. "The Place of Plants: Spatiality, Movement, Growth." *Performance Philosophy* 1 (April 2015), 185–74.

Margolis, Joseph. "What, after All, Is a Work of Art?". In *What, after All, Is a Work of Art?: Lectures in the Philosophy of Art*. University Park, PA: Penn State Press, 1999.

"Masters of the 20th Century." <https://www.kunsthauswien.com/en/exhibitions/archive/38-2000/136-tapisserie>.

Matthews, Richard. *To Whom It May Concern: A Romp through the Life of Richard Matthews*. New Zealand: R. Matthews, 2009.

Meskin, Aaron. "From Defining Art to Defining the Individual Arts: The Role of Theory in the Philosophies of Arts." In *New Waves in Aesthetics*, eds Kathleen Stock and Katherine Thomson-Jones. Houndmills, Basingstoke, Hampshire and New York: Palgrave Macmillan, 2007.

Miller, Mara. *The Garden as an Art*. Albany: SUNY Press, 1993.

——. "Gardens as Works of Art: The Problem of Uniqueness." *British Journal of Aesthetics* 26, no. 3 (Summer, 1986). 252–6.

——. "Gardens: Gardens as Art." In *Encyclopedia of Aesthetics*, ed. Michael Kelly. New York, Oxford: Oxford University Press, 1998.

——. "Time and Temporality in the Garden." In *Gardening: Cultivating Wisdom*, ed. Dan O'Brien. Oxford: Wiley-Blackwell, 2010.

Muir, John. "The Wild Parks and Forest Reservations of the West." *The Atlantic Monthly* 81, no. 483 (1898). Published electronically. <http://vault.sierraclub.org/john_muir_exhibit/writings/favorite_quotations.aspx>.

Myers, Katherine. "Visual Fields: Theories of Perception and the Landscape Garden." In *Experiencing the Garden in the Eighteenth Century*, ed. Martin Calder. Oxford; New York: Peter Lang, 2006.

Naravane, Vishwanath S. *Ananda K. Coomaraswamy*. Twayne World Leader Series. Vol. 75, Boston, MA: Twayne Publishers, 1978.

Noggle, James. *The Temporality of Taste in Eighteenth-Century British Writing*. Oxford: Oxford University Press, 2012.

Novack, Cynthia J. *Sharing the Dance: Contact Improvisation and American Culture*. Madison: University of Wisconsin Press, 1990.

Parshall, Linda. "Motion and Emotion in C. C. L. Hirschfeld's Theory of Garden Art." In *Landscape Design and the Experience of Motion*, ed. M. Conan. Washington, DC: Dumbarton Oaks Research Library and Collection, 2003.

Parsons, G. "The Aesthetics of Nature." *Philosophy Compass* 2, no. 3 (2007), 358–72.
Parsons, Glenn. *Bloomsbury Aesthetics: Aesthetics and Nature*. London: Continuum, 2008.
Paxton, Steve. "Contact Improvisation." *The Drama Review* 19 (1975), 40.
Plato. "The Republic: Book X."
Pope, Alexander. *An Essay on Criticism*. <https://www.gutenberg.org/files/7409/7409-h/7409-h.htm>.
Powell, John. "Music, Gardens, and Time." In *Time Theories and Music*. Department of Music Studies, Ionian University, Corfu, 2012.
——. "We Do Not Have an Adequate Conception of Art until We Have One That Accommodates Gardens." MA, Victoria University of Wellington, 2012.
——. "What Is Temporal Art? A Persistent Question Revisited." *Contemporary Aesthetics* 13 (2015).
Price, Uvedale. *Essays on the Picturesque, as Compared with the Sublime and the Beautiful; and, on the Use of Studying Pictures, for the Purpose of Improving Real Landscape*. 2 vols. [Electronic resource.] London: J Mawman, 1810.
Rainer, Yvonne. *Work 1961–73*. Halifax, Nova Scotia: Press of the Nova Scotia College of Art and Design, 1974.
Rosenberg, Harold. *The Anxious Object: Art Today and Its Audience*. New York: Horizon Press, 1964.
Ross, Stephanie. "Gardens' Powers." *Journal of Aesthetic Education* 33, no. 3 (1999).
——. "Paradoxes and Puzzles: Appreciating Gardens and Urban Nature." *Contemporary Aesthetics* 4 (2006), 1–13.
——. *What Gardens Mean*. Chicago: University of Chicago Press, 1998.
Ruggles, D. Fairchild, ed. *Sound and Scent in the Garden*, Dumbarton Oaks Colloquium on the History of Landscape Architecture, vol. 38. Harvard University Press, 2017.
St-Denis, Bernard. "Just What Is a Garden?" *Studies in the History of Gardens & Designed Landscapes* 27, no. 1 (2007), 61–76.
Saito, Yuriko. *Everyday Aesthetics*. New York: Oxford University Press, 2008.
Salwa, Mateusz. "The Garden as a Performance." *Proceedings of the European Society for Aesthetics* 5 (2013), 372–87.
Scruton, Roger. *Perictione in Colophon: Reflections on the Aesthetic Way of Life*. South Bend, IN: St Augustine's Press, 2000.
Shapshay, Sandra. "Schopenhauer's Aesthetics." In *The Stanford Encyclopedia of Philosophy*, ed. E. N. Zalta. Summer 2012. <http://plato.stanford.edu/archives/sum2012/entries/schopenhauer-aesthetics/>.
Shelley, James. "18th-Century British Aesthetics." In *The Stanford Encyclopedia of Philosophy*, ed. Edward N. Zalta. Fall 2014 edn. <http://plato.stanford.edu/archives/fall2014/entries/aesthetics-18th-british/>.

——. "The Concept of the Aesthetic." In *The Stanford Encyclopedia of Philosophy*, ed. Edward N. Zalta. Winter 2015. <http://plato.stanford.edu/archives/win2015/entries/aesthetic-concept>.

Shiner, Larry. "Art Scents: Perfume, Design and Olfactory Art." *British Journal Of Aesthetics* 55, no. 3 (July 2015), 375–92.

Shusterman, R. "The End of Aesthetic Experience." *The Journal of Aesthetics and Art Criticism* 55, no. 1 (1997), 29–41.

Shusterman, Richard. *Body Consciousness: A Philosophy of Mindfulness and Somaesthetics*. New York: Cambridge University Press, 2008.

"Sound and Scent in the Garden: Garden and Landscape Studies Symposium at Dumbarton Oaks." <http://www.doaks.org/research/garden-landscape/scholarly-activities/past/sound-and-scent-in-the-garden/abstracts>.

Spackman, John. "Expression Theory of Art." In *Encyclopedia of Aesthetics*, ed. Michael Kelly, Oxford University Press, 2014. <http://www.oxfordreference.com/view/10.1093/acref/9780195113075.001.0001/acref-9780195113075-e-0204>.

"Special Issue on Time." *Studies in the History of Gardens & Designed Landscapes: An International Quarterly* 34, no. 1 (2014).

Tafalla, Marta. "Anosmic Aesthetics." *Estetika: The Central European Journal of Aesthetics* 50, no. 1 (2013), 53–80.

——. "Smell and Anosmia in the Appreciation of Gardens." *Contemporary Aesthetics* 12 (2014).

Taranaki Regional Council. "Tupare: Relive the Splendour." Taranaki Regional Council, <http://www.trc.govt.nz/tupare-home/>.

"Temporality, N." In *OED Online*. Oxford University Press. <http://www.oed.com/view/Entry/198948?redirectedFrom=temporality>.

Thomasson, Amie L. "Debates About the Ontology of Art: What Are We Doing Here?". *Philosophy Compass* 1, no. 3 (2006), 245–55.

——. "Ontological Innovation in Art." *The Journal of Aesthetics and Art Criticism* 68, no. 2 (2010), 119–30.

——. "The Ontology of Art." In *Blackwell Guide to Aesthetics*, ed. Peter Kivy. Hoboken, NJ: Wiley-Blackwell, 2009.

——. "The Ontology of Art and Knowledge in Aesthetics." *The Journal of Aesthetics and Art Ciriticsm* 63, no. 3 (2005), 221–9.

Turner, Tom. *Garden History: Philosophy and Design 2000 B.C.–2000 A.D.* London and New York: Spon Press, 2005.

Walton, Kendall "Categories of Art." *The Philosophical Review* 79, no. 3 (1970), 334–67.

Warhol, Andy. "*Brillo Boxes*." New York: Stable Gallery, 1964.

Weitz, Morris. "The Role of Theory in Aesthetics." *The Journal of Aesthetics and Art Criticism* 15, no. 1 (1956), 27–35.

Williams, Raymond. *Keywords: A Vocabulary of Culture and Society*. New York: Oxford University Press, 1976.

Wimsatt, W. K., Jr, and M. C. Beardsley. "The Intentional Fallacy." *The Sewanee Review* 54, no. 3 (1946), 468–88.

Wollheim, Richard. *Art and Its Objects*. 2nd edn. Cambridge: Cambridge University Press, 1980.

——. *Art and Its Objects: An Introduction to Aesthetics*. New York: Harper and Row, 1968.

Young, Damon. "Philosophy Alfresco." In *Voltaire's Vine and Other Philosophies: How Gardens Inspired Great Writers*. London: Ebury Digital, 2014.

Index

Cultural Interactions
Studies in the Relationship between the Arts

Edited by J.B. Bullen

Interdisciplinary activity is now a major feature of academic work in all fields. The traditional borders between the arts have been eroded to reveal new connections and create new links between art forms. Cultural Interactions is intended to provide a forum for this activity. It will publish monographs, edited collections and volumes of primary material on points of crossover such as those between literature and the visual arts or photography and fiction, music and theatre, sculpture and historiography. It will engage with book illustration, the manipulation of typography as an art form, or the 'double work' of poetry and painting and will offer the opportunity to broaden the field into wider and less charted areas. It will deal with modes of representation that cross the physiological boundaries of sight, hearing and touch and examine the placing of these modes within their representative cultures. It will offer an opportunity to publish on the crosscurrents of nationality and the transformations brought about by foreign art forms impinging upon others. The interface between the arts knows no boundaries of time or geography, history or theory.

Vol. 1 Laura Colombino: Ford Madox Ford: Vision, Visuality and Writing
275 pages. 2008. ISBN 978-3-03911-396-5

Vol. 2 Graham Smith: 'Light that Dances in the Mind': Photographs and Memory in the Writings of E. M. Forster and his Contemporaries
257 pages. 2007. ISBN 978-3-03911-117-6

Vol. 3 G.F. Mitrano and Eric Jarosinski (eds): The Hand of the Interpreter: Essays on Meaning after Theory
370 pages. 2009. ISBN 978-3-03911-118-3

www.ingramcontent.com/pod-product-compliance
Ingram Content Group UK Ltd.
Pitfield, Milton Keynes, MK11 3LW, UK
UKHW020719041026
15590UKWH00002B/754

9 781789 971415